HAWKEYE INST. OF TECHNOLOGY
3 7944 1001 4797 0

D0073048

PRACTICAL RF DESIGN MANUAL

Doug DeMaw

Member, IEEE, Graduate Engineer

Prentice-Hall, Inc., *Englewood Cliffs, N. J. 07632*

Library of Congress Cataloging in Publication Data

DeMaw Doug. (date)
Practical RF design manual.

Bibliography: p.
Includes index.
1. Radio circuits. 2. Electronic circuit design.
I. Title.
TK6553. D342 621.3841'2 81-14378
ISBN 0-13-693754-3 AACR2

Editorial/production supervision and interior design: Nancy Moskowitz
Manufacturing buyer: Gordon Osbourne
Cover Design: Edsal Enterprise

Printed in the United States of America

10 9 8 7 6 5 4 3 2 1

ISBN 0-13-693754-3

Prentice-Hall International, Inc., *London*
Prentice-Hall of Australia Pty. Limited, *Sydney*
Prentice-Hall of Canada, Ltd., *Toronto*
Prentice-Hall of India Private Limited, *New Delhi*
Prentice-Hall of Japan, Inc., *Tokyo*
Prentice-Hall of Southeast Asia Pte. Ltd., *Singapore*
Whitehall Books Limited, *Wellington, New Zealand*

CONTENTS

PREFACE

This is the communications area. If we are to communicate in a lucid and beneficial manner, it is important to avoid unnecessary engineering vernacular, page-long equations, and superficial treatment of the subjects with which we deal. This book does not contain the lofty language found in many of today's texts. Rather, plain language is used throughout, without "talking down" to the reader. Equations are used only where it is necessary to demonstrate a particular design approach. Approximations are used where acceptable, owing to the nonuniformity found in semiconductors that are not graded for a tight set of parameters. It is usually required that the designer do some empirical finalizing of a circuit to compensate for the differences in performance of transistors or ICs that bear the same part number. Therefore, the most concise mathematical design effort will seldom yield the performance objective set by the engineer or student.

The circuits in this volume are founded on practical experience in the laboratory. The reader will find no theoretical circuit examples, and there is little reference to circuits that have appeared in trade journals and manufacturers' application notes. Although the circuits provided in this book are not intended for exact duplication, they will function well if they are built. Many of them can be used as building blocks for composite systems in the rf communications industry.

The author wishes to acknowledge the unforseen contributions made by his professional colleagues, who through numerous circuit discussions stimulated his thinking respective to better techniques in rf circuit design. Special recognition belongs to Wes Hayward of Tektronics, Merle Hoover of RCA, Ed Oxner of Siliconix, Dr. Ulrich Rohde of Rohde & Schwarz, and George Woodward of the ARRL, Inc., staff. Our frequent discussions and debates were inspiring and fruitful.

The author dedicates this book to his son, Douglas David DeMaw, who has just begun his professional career as a technical writer.

Doug DeMaw

1

TRANSMITTER AND RECEIVER FUNDAMENTALS

In this chapter we will examine the basic considerations of transmitter and receiver performance. Generally, the concepts treated in this chapter will serve as the necessary guidelines which apply to the discussions that are contained in the remainder of this volume. To ensure the execution of a quality design, it is important that we understand what our objectives must be. It is vital to set specific criteria that will lead to acceptable modern-day circuit performance for the commercial broadcasting frequencies from VLF through UHF. Therefore, it is in your better interests to read this chapter before moving ahead with your study of the detailed information that follows.

1.1 TUBES VERSUS SEMICONDUCTORS

Although we are well committed at present to the use of semiconductor devices (transistors, diodes, integrated circuits, and LEDs) for small-signal applications, vacuum tubes are still being used in high RF-power circuits. The decision to employ tubes rather than power transistors is founded on practical considerations for the most part. As circuit designers, we must weigh one aspect of the chosen active device against that of its counterpart: tubes versus semiconductors. Let's examine the major considerations in the order of their significance.

1. Total cost, inclusive of power supplies.
2. Efficiency.

3. Physical size of production model.
4. Weight of the assembled product.
5. Longevity and reliability.
6. Performance integrity.
7. Ease of maintenance.
8. Adaptability to rapid-production techniques.

Although there are other characteristics to contemplate when starting a design effort, the foregoing list represents our area of primary concern.

1.2 DECISION MAKING

Let us examine the list in Section 1.1. Item 1 is based entirely on economics. For the purpose of discussion, let's imagine that our design objective is to develop a 500-watt (W) RF power amplifier that operates in the linear mode at 25 megahertz (MHz). A good tube choice would be a 3-500Z power triode. Figure 1.1 shows a typical circuit in which the 3-500Z tube might be employed. If we compare the solid-state circuit equivalent of Figure 1.2 with that of Figure 1.1, it becomes readily apparent that the transistorized amplifier requires many more components than the tube version to provide the same output power. Therefore, the cost of component parts is likely to be substantially lower when designing for the 3-500Z tube. Additionally, the cost of four 150-W transistors will be somewhat greater than for a single 3-500Z.

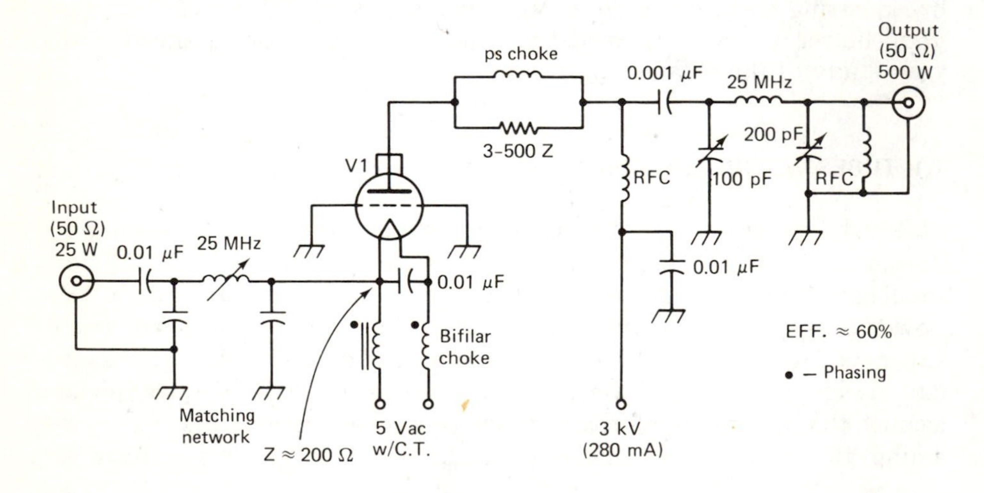

Figure 1.1 Circuit for a 500-W power amplifier that uses a 3-500Z vacuum tube.

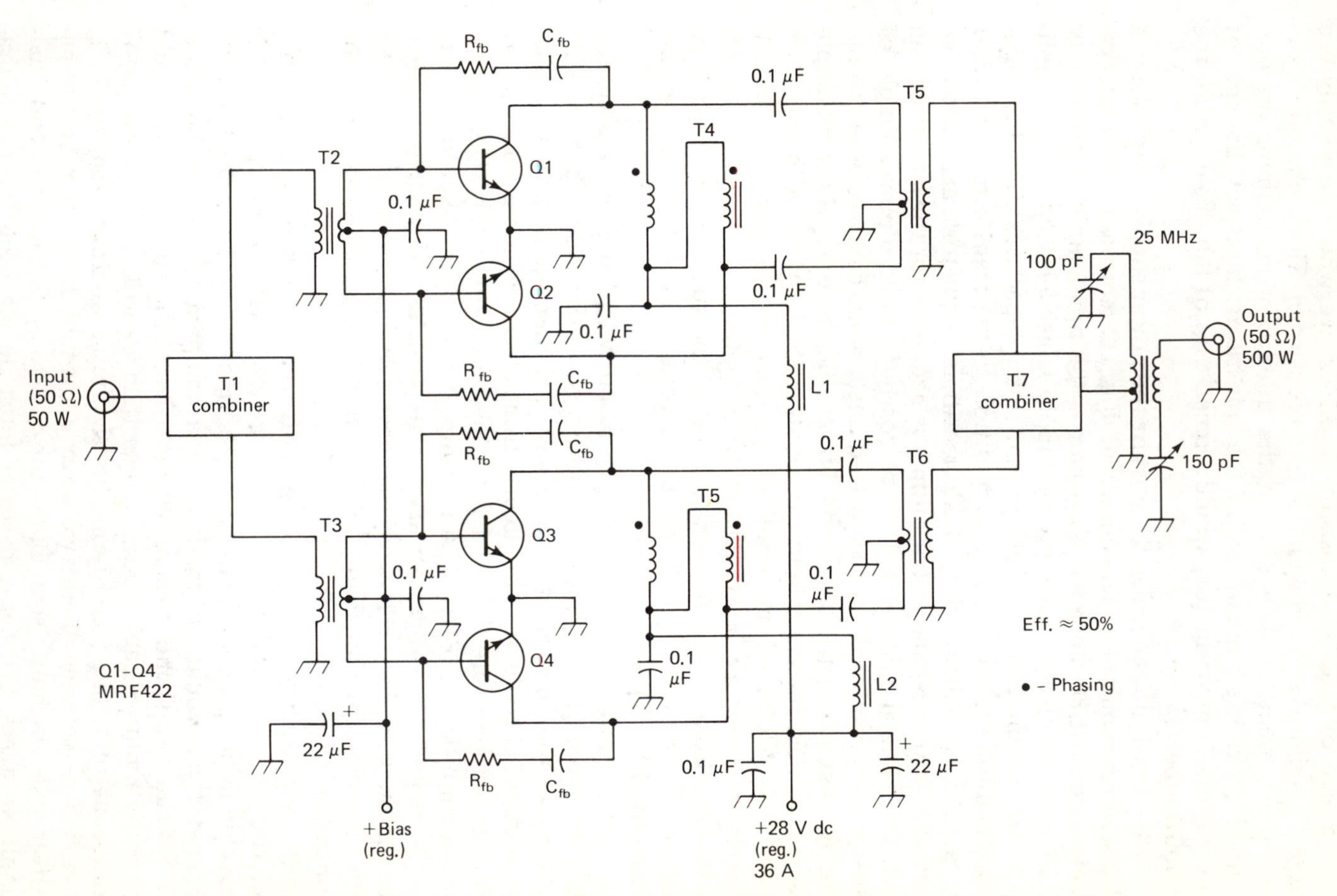

Figure 1.2 High-power 500-W RF amplifier using transistors. Two push-pull amplifiers have their outputs added by means of a combiner, *T7*.

Forced-air cooling will be required with either style of amplifier. In the case of the four-transistor amplifier, we would necessarily use large finned heat sinks in addition to a blower fan. This technique is the rule rather than the exception in the interest of maintaining the prescribed operating temperature of the transistors. In effect, the cost of cooling the semiconductors would exceed that of the single fan for the 3-500Z tube.

We must also consider the cost of the power supply when basing a decision on economics. The circuit of Fig. 1.2 requires an operating direct current of 36 amperes (A), whereas the tube amplifier needs only 0.28 A. A relatively simple power supply is needed for the tube amplifier (transformer, bridge rectifier, and filter capacitors), but the solid-state amplifier would require, in addition to the transformer and filter capacitors, some heavy-duty diode rectifiers, regulators, and protective circuits. Also, a regulated bias supply is necessary for the circuit of Fig. 1.2. In the final analysis, it is more costly to build a power supply of the type needed for the transistorized linear amplifier as opposed to the one we would construct for the 3-500Z tube. In terms of overall amplifier efficiency, the two circuits compare rather closely. Although a typical class AB solid-state amplifier exhibits an efficiency of approximately 50%, and a class B tube amplifier approaches or slightly exceeds 60% in efficiency, the filament power of the tube amplifier tends to balance the scale. A 3-500Z requires 5 volts (V) at 14.5 A to light the filament (72.5 W).

At the lower power levels, such as 50 W, the solid-state amplifier may be more practical in terms of unit cost. This would be especially true if we were designing the equipment for mobile service: the solid-state amplifier could operate directly from the 12-V battery of a vehicle. This would eliminate the need for a special power supply. Conversely, a tube amplifier (6146B tube, for example) of comparable output power would require a dc-to-dc converter to step the 12-V dc source to 600 V dc for the tube anode. It would be necessary also to obtain 6.3 V for the tube filament, 250 V for the screen grid, and 50 V negative for the control-grid bias.

Since we have already discussed efficiency, we will move to item 3 on the list of considerations, *physical size*. Without exception, it is possible to build a composite solid-state transmitter into a much smaller cabinet than would be possible for a comparable all-tube circuit. A 500-W single-sideband transmitter, inclusive of the power supply, would typically require a cabinet much larger than we would need for a 500-W solid-state unit. In general terms the size ratio would be roughly 3 : 1. Somewhere between these two sizes we would find the hybrid transmitter, one which uses semiconductor devices up to the driver and final-amplifier stages. Thus, the designer must decide what the size limitations will be before starting the design procedure.

Item 4 of the list of considerations relates to *weight*. For non-mobile use, it is likely that solid-state and tube transmitters of like power will have similar weights. This assumes that the ac-operated power supplies are included in the package. In a practical situation we would not be constricted by size limits unless the product was being designed for aeronautical, shipboard, or land-mobile service.

Longevity and reliability, as referenced in item 5 of our list, open the door to considerable debate among tube and semiconductor design engineers. Each group will opt for its favorite amplifying device. But if we are to be totally objective in this discussion, we will strongly favor the semiconductor over the vacuum tube. The assumption is that the circuit is designed and operated within the constraints of good engineering practice. Tubes are fragile, comparatively speaking. They suffer from long-term degradation of emission, can become gassy, and are subject to internal short-circuiting. None of these maladies befalls transistors if they are operated within their ratings.

It should be said that tubes are far more immune to catastrophic failure than transistors are. Momentary excesses of operating voltage and current seldom damage a vacuum tube. Conversely, most transistors will be destroyed immediately in the presence of excessive voltage and current. Therefore, it becomes necessary to include certain protective circuits in semiconductor power circuits to ensure the longevity of the devices. The life-span of a transistor or integrated circuit (IC) should exceed the expected or intended existence of the piece of equipment in which it is used. This cannot be said of vacuum tubes if the equipment is subjected to regular use.

We look next to the matter of *performance integrity*, which is the sixth item in our list of considerations. We must be concerned in the case of a transmitter that the spectral purity of the output waveform complies with the standards of good engineering practice. This means that harmonics and other spurious components should comply with FCC rules and regulations. In this context, there should be no spurious energy that is greater in amplitude than -40 decibels (dB), as referenced to peak output power in the HF spectrum. The requirement calls for -60 dB in the VHF portion of the spectrum.

In making a performance and design comparison between vacuum-tube transmitters and those that employ semiconductor amplifiers, the tube has the clear advantage with respect to harmonic output currents. Conventional *envelope distortion* is the principal cause of harmonic energy in tube types of amplifiers. Transistor amplifiers, on the other hand, generate harmonic currents through the function of envelope distortion, but develop even greater magnitudes of harmonic current by virtue of nonlinear changes in junction capacitance. This phenomenon is brought about as the sine wave swings through its normal excursion. The effect is generally the same as that which results from frequency multiplication by means of voltage-variable capacitor diodes (varactors). In a practical solid-state RF amplifier, the second and third harmonics of the desired signal are only 10 to 15 dB lower in amplitude (at the transistor collector) than the fundamental signal.

In our design efforts we must consider the effects of this unusually high level of harmonic current and be aware of its effects on overall performance. For example, excessive harmonic energy in the low-level stages of a transmitter can cause the successive transmitter stage to be driven too hard with respect to the output level sought at the desired frequency. Not only is this wasteful of dc power, it can cause troublesome mixing within the driven stage, thereby producing unwanted sum-and-difference frequencies in the output of the driven stage. It is by no means unusual to find numerous strong spurious products below the operating frequency of a solid-state transmitter which has not been designed carefully. If, for example, the transmitter was designed for operation at 10 MHz, spurs could be found below 10 MHz, perhaps extending down into the VLF range. This condition would be hard for us to duplicate while using vacuum tubes.

In view of the foregoing causes and effects, it is essential that we, as designers, direct our attention toward minimizing spurious energy in a solid-state transmitter. Because of this, it becomes the rule rather than the exception to include some fairly elaborate filter networks in solid-state transmitters. These measures are seldom necessary in an equivalent tube type of transmitter. Figure 1.3 shows a general comparison between a single-device RF amplifier in which we might utilize a tube or a transistor. The tube version of Fig. 1.3(a) requires only the high-Q tuned circuits at the grid and plate to provide effective attenuation of the close-in harmonics. It is not uncommon, however, to find a low-pass filter used at the output of this amplifier for the purpose of attenuating VHF harmonic energy. For example, if the amplifier were designed to be a band-switched variety, with coverage from 1.5 to 30 MHz, the filter cutoff frequency would be approximately 40 MHz.

As we examine the circuit of Fig. 1.3(b), we observe two harmonic filters (FL1 and FL2) that are half-wave low-pass networks. In this example, they have been designed for a loaded Q of 1 at 50 ohms. $T1$ and $T2$ serve as broadband matching transformers to permit interfacing the low-impedance base and collector ports to the 50-Ω harmonic filters. Some designers incorporate a bypass capacitor ($C1$) in shunt with the collector to remove VHF harmonic energy. If $C1$ is used, it should exhibit high reactance at the operating frequency.

We can conclude from the examples in Figure 1.3 that performance integrity can require considerably more engineering effort when transistors are used in RF power amplifiers. The design exercise is not especially difficult, however, because the application of broadband transformers and filters is no more esoteric than is the proper design

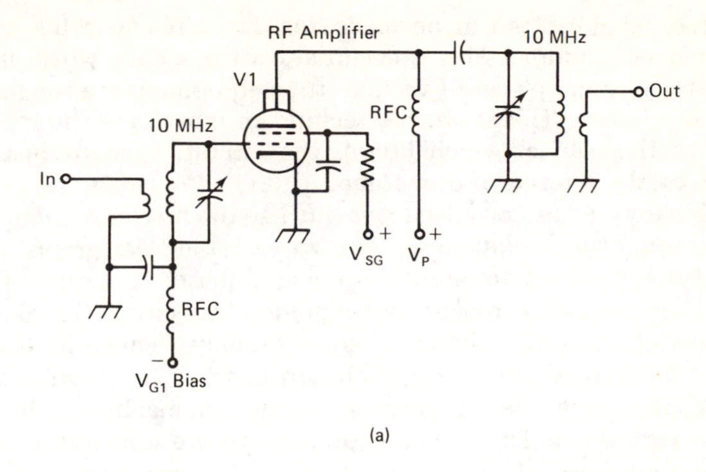

(a)

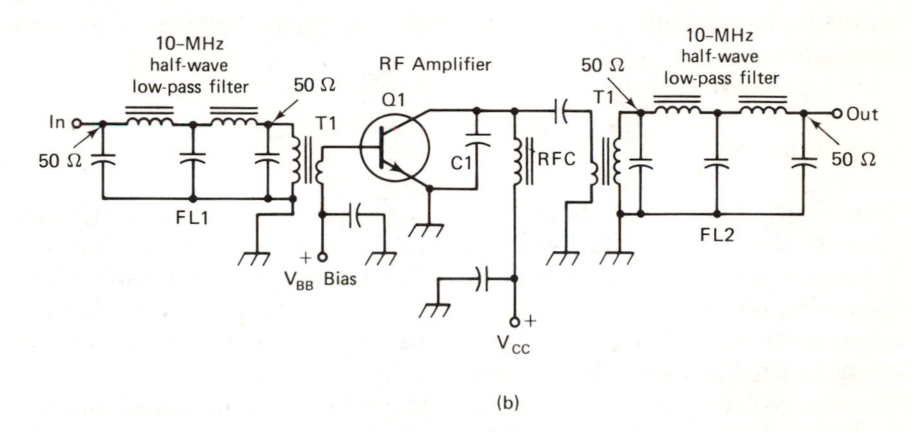

(b)

Figure 1.3 Comparison between (a) a power type of tube amplifier, and (b) a solid-state equivalent.

and use of high-Q tuned circuits for vacuum-tube amplifiers. In both situations we must not only reduce the harmonic output, but ensure an effective impedance match between the tube and its source and load.

We find *ease of maintenance* next on our list of points to heed when chosing between tubes and semiconductors. In this regard it is important to be aware that most solid-state circuits are built on printed-circuit boards. This is seldom the case with vacuum-tube circuit modules. The advantage of circuit boards can be seen when making repairs or circuit updates. A plug-in circuit-board module can be removed quickly and replaced with a working unit. This expediency is particularly important in the broadcast industry and where military equipment is involved. The "downtime" is minimal as compared to that of a composite tube-type piece of equipment.

Extender cables can be made for use when working on a "live" circuit-board module. This is essential during service work on a bench. Conversely, some pieces of vacuum-tube equipment are too large to service on a bench. Therefore, the technician must take the repair equipment to the cabinet which houses the circuit. In many instances this can be costly in terms of downtime.

Finally, let us consider from our list the matter of *adaptability to rapid production techniques*. The two foregoing paragraphs fairly well identify the solid-state circuit as being superior in terms of handling ease. Therefore, we can rightfully conclude that circuit-board assemblies are capable of being fabricated more quickly than units that contain tubes. A complex printed-circuit board can have its parts mounted by means of a computer-programmed insertion machine. The insertion time is very short. Then all the components are soldered to the circuit-board conductors at one time. This is done by means of a dip-solder technique or a wave soldering machine. Custom point-to-point soldering would be required of a similar vacuum-tube circuit that was built on a metal chassis.

1.3 RECEIVERS

Most of what we have discussed thus far has been keyed to transmitter circuits. But the practical comparisons between tubes and semiconductors apply equally when beginning the design of receiving apparatus. Unit cost, physical size, longevity, ease of maintenance, and production expediency are just as vital to the manufacturer and engineering staff as when transmitters are being fabricated.

A myth has existed for many years among proponents of vacuum-tube equipment. It concerns the unfounded belief that transistors are noisier and more fragile than tubes are. We must not allow ourselves to succumb to this false doctrine. The fact of the matter is that in most small-signal applications the transistor or diode is equal to or better than the vacuum tube. In the chapters that follow we will approach the design and application subject in a positive, contemporary manner: tubes will be considered only where they provide better performance with respect to maximum obtainable power levels versus manufacturing and maintenance cost.

1.4 TRANSMITTER PERFORMANCE

There are some succinct performance characteristics we must perceive as guidelines when designing a piece of transmitting equipment. Unfortunately, there are no published engineering standards for some of the parameters we must concern ourselves with. The criteria that follow are offered as recommendations only. The eight points we are highlighting emerge from many years of laboratory experience and design connected with the development of HF and VHF transmitting equipment on behalf of the author. The criteria are founded in part on observations made while testing various pieces of commercial transmitting apparatus, some of which complied with military specifications. Each of the eight points, plus others, will be dealt with in considerable detail later in this book.

1.4.1 Frequency Stability

An ideal transmitter would exhibit no measurable frequency drift from the time it was turned on to some point much later in time—days, weeks, or even months. In a practical situation there is seldom a need for stability of this magnitude. Rather, certain tolerable limits are acceptable to most services. Transmitters that utilize frequency synthesizers are among the most stable types today. Crystal-controlled transmitters can be made as stable as those which use frequency synthesis, provided the crystal is contained in a crystal oven of suitable characteristics. The least stable type of frequency-controlling circuit is the *LC* oscillator or variable-frequency oscillator (VFO). All the circuits discussed in this section are the most stable when semiconductors are used rather than vacuum tubes. This is because the tube type of transmitter or oscillator generates a substantial amount of heat. This heat has a pronounced effect on the long-term stability of the transmitter. Although temperature-compensating elements, such as negative, positive, or zero temperature coefficient capacitors, can be employed as a corrective measure for frequency drift brought on by increases in ambient temperature, the net gain is not always satisfactory. Much better results are obtained when semiconductor devices are employed in the equipment.

We are concerned with two classes of drift: short term and long term. Short term drift takes place during the first few minutes of operation, while the tube elements or semiconductor junctions rise to their normal operating temperatures. Drift of this variety is of least concern to the designer or operator, since all equipment must be subjected to a specified normal warm-up period when placed in service.

Long-term stability is a matter of far greater concern than is short-term drift. Excessive frequency change with time requires the operator to make corrections frequently. Not only is this an inconvenience, it can have a marked effect on the integrity of the transmitting station when important communications are taking place: if the transmitted signal is far enough off the prescribed frequency, the intelligence of the communication can be impaired sufficiently to render it unreadable. Furthermore, the licensee of the transmitter may receive a citation from the FCC for being in violation of his or her license provisions.

Long-term frequency shift is caused by a steady change in ambient temperature in the area where the transmitter oscillator is located. Additional to the temperature increase caused within the equipment cabinet as the various active devices generate heat is the change caused by the external environment. Examples of marked changes in temperature outside the equipment enclosure are seen in the case of land-mobile and airborne applications. Operation in these environments can commence at temperatures well below freezing. Once the vehicle is placed in service, its heating system can subject the transmitter to very high temperatures, depending upon where the equipment is installed in the vehicle. This external temperature change, with time, is a significant part of the long-term-drift period.

To compensate for the kinds of drift just discussed, we must understand why heat has such a profound effect on frequency stability. Such components as fixed-value capacitors, variable capacitors, and magnetic core materials (ferrite or powdered iron) undergo changes in characteristics as their temperatures change. Depending upon the material used in these components, the drift in value with heat can be positive or negative. The characteristic must be known in order for the designer to take the proper compensatory steps toward stabilization. The component manufacturer can provide the designer with a stability profile of a product, if that information is desired.

Other significant causes of temperature-induced drift are RF circulating currents within the capacitors and inductors of an oscillator circuit. The circulating currents, no matter how miniscule they might be, cause internal heating of fixed-value capacitors. These currents also heat the windings of inductors to some extent, and this can cause minor expansion of the wire. Such changes will have an effect on the *effective inductance* of the coil by virtue of changes in distributed capacitance across the coil turns. If a magnetic core is used in the coil, its properties (permeability) can undergo minor changes because of RF current. Generally, this type of change is *positive* in nature, causing some *increase* in inductance.

The undesirable effects of RF circulating currents can be reduced by increasing the area of the capacitor plates. This is most easily achieved by using capacitors in parallel in critical parts of the circuit. A comparison between the basic VFO and one that has been designed for improved stability through the use of paralleled fixed-value capacitors is seen in Fig. 1.4(a) and (b). The significant capacitors are $C2$, $C3$, and $C4$ of Fig. 1.4.(a). The variable capacitor, $C1$, normally has sufficient mass to let us ignore it with respect to heating versus circulating current.

Examination of the revised circuit shown in Fig. 1.4(b) reveals the addition of three fixed-value capacitors. $C2$, $C3$, and $C4$ now consist of paralleled pairs of capacitors, as opposed to the single condition of Fig.

1.4(a). The net capacitance values are the same in both circuits, but each point in the frequency-determining part of the circuit has twice the current-carrying area. This greatly minimizes heating within a single unit, and hence reduces the change in capacitance that can be caused by circulating current. The 500-picofarad (pF) gate coupling capacitor could be treated in the same manner, but in a practical circuit the effects of RF current are minimal in that part of the oscillator. The same applies to the 50-pF output coupling capacitor.

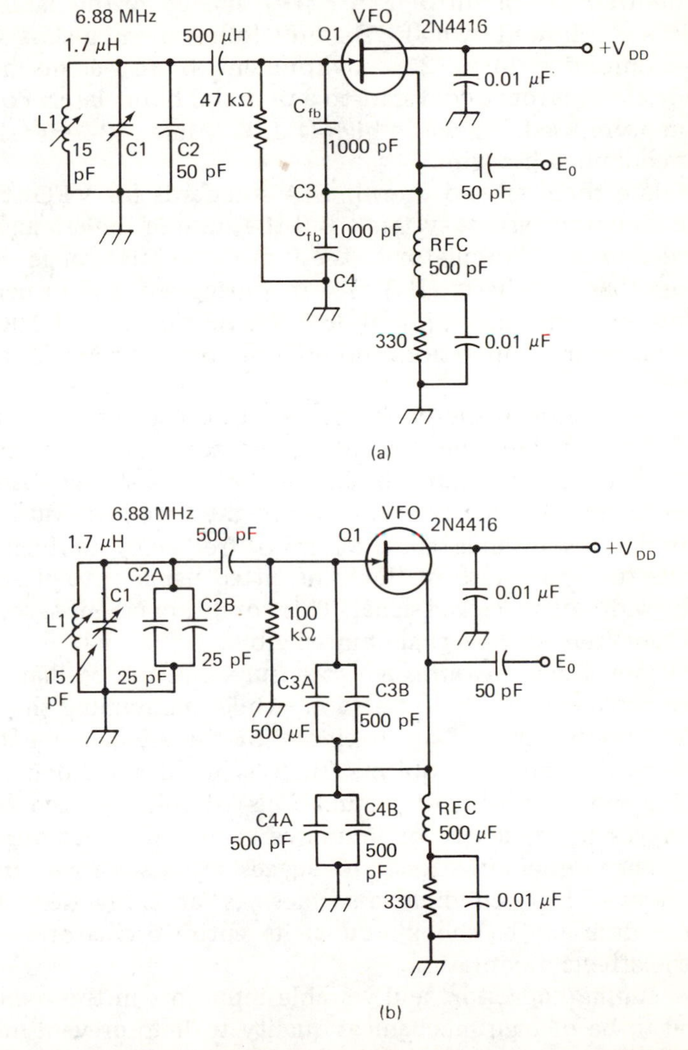

Figure 1.4 (a) Basic Colpitts type of VFO; (b) circuit using additional frequency-determining capacitors (parallel) to reduce drift effects from RF heating within the capacitors.

If $L1$ of Fig. 1.4 contains a magnetic-core slug, there will be a positive drift condition for the inductor. Therefore, it is necessary to introduce negative compensation. This can be most easily done by utilizing negative-coefficient capacitors at $C2$, $C3$, and $C4$. Some designers prefer NPO ceramic capacitors for this purpose. Excellent results can also be obtained by using polystyrene capacitors at $C2$, $C3$, and $C4$. They are very temperature stable and have a negative-coefficient trait. Polystyrene capacitors should work satisfactorily up to at least 10 MHz. The author designed a series-tuned Colpitts VFO for use from 2.0 to 2.2 MHz, which followed the circuit of Fig. 1.4(b), except for the parallel tuning of $L1$. It was followed by a JFET source follower and a class A bipolar-transistor amplifier. In a 72°F environment, there was no measurable frequency change from a cold start to a period 3 hours later. Polystyrene capacitors were used. $L1$ was a high-Q J. W. Miller 43-series slug-tuned coil that contained litz wire.

To date there are no known EIA standards for VFO stability in HF-band communications systems. But the state of the art suggests that no well-designed LC oscillator in the 0.5- to 10-MHz range should exhibit more than 100 hertz (Hz) of drift during any 1-hour period after initial turn-on. Practically, initial drift should not exceed 100 Hz during the first hour of operation, and no more than ±25 Hz for each hour thereafter.

Another factor in the performance of LC oscillators is *mechanical stability*. This concerns the oscillator characteristics with respect to vibration. An otherwise stable oscillator can generate sidebands when subjected to stress of the type that we might encounter during mobile operation. These sidebands take the form of frequency modulation (FM) to increase the bandwidth of the transmitted signal or to effectively increase the width of a received signal. This condition can, if severe enough in magnitude, render the signal unintelligible.

The preventive measures are to ensure that the oscillator components are affixed securely in position, while maintaining short pigtails and other circuit leads. The proximity of the frequency-determining components to conductive objects (such as metal shield plates, variable capacitors, and circuit-board ground foils) should be such that minimum changes in capacitive or inductive value will result during vibration or other mechanical stress. This suggests that sensitive components be spaced away from conductive surfaces as far as practical. Also, it is sometimes desirable to shock-mount the entire oscillator assembly to reduce the effects of vibration.

The tuning capacitors and variable inductors in the oscillator circuit need to be of high mechanical quality to help prevent mechanical instability. The capacitors should be of the double-bearing variety so that the rotor is supported at each end. The collets of slug-tuned inductors require similar mechanical integrity to minimize unwanted movement during periods of mechanical stress.

1.4.2 Spectral Purity

The RF energy that emerges from the output port of a transmitter should consist primarily of the desired output signal or *fundamental frequency*. This truism could be realized in only an ideal transmitter. Therefore, we must attempt to approach the ideal as closely as possible when designing and perfecting a piece of transmitting equipment.

Perhaps the first priority is complying with government regulations that pertain to the purity of emissions. FCC type approval and type acceptance depend largely on how well the manufacturer has suppressed both the in-band and out-of-band spurious responses of a transmitter. Specific minimum acceptable spurious levels exist for various modes of transmission in various parts of the radio spectrum. Since these regulations are subject to change at any time, the suppression levels versus modes and frequencies will not be stated here. Such information is readily available from the FCC in Washington, D.C.

The second priority we must consider in designing a piece of transmitting equipment is the matter of *good engineering practice*. It is not sufficient to merely "comply" with the regulations for spectral purity. The present state of the art is such that it is not difficult or expensive to exceed the minimum requirements that have been imposed. As we study the following chapters, we will learn how all of this is possible. The second priority might be summed up in two words: *engineering pride*.

The basic causes of noncompliance with the regulations for purity of emissions are excessive harmonic output, spurious mixing products, and unwanted self-oscillation in one or more stages of a transmitter. Self-oscillation may become manifest at VHF, UHF, HF, MF, or LF, irrespective of the operating frequency of the transmitter. Some solid-state HF and VHF transmitters are prone to self-oscillations as low as the audio-frequency range. This is one of the principal causes of in-band spurs.

Let us assume that the regulations for a particular type of HF-band transmitter called for suppression of the even- and odd-order harmonics to be –40 dB or greater. The designer chose to cut the product expense by minimizing the number of poles in his or her output filter. The spectral display might therefore look like that shown graphically in Fig. 1.5(a). The second harmonic is down only 25 dB from peak power. The third harmonic falls short of compliance also, since it is down 35 dB from peak output power. The fourth, fifth and sixth harmonics are

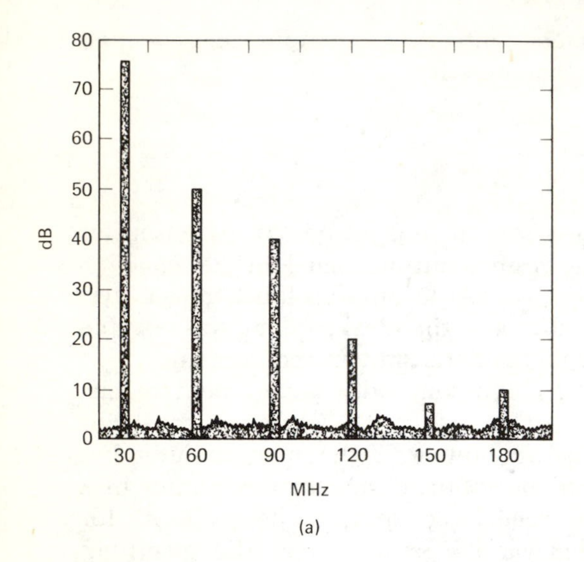

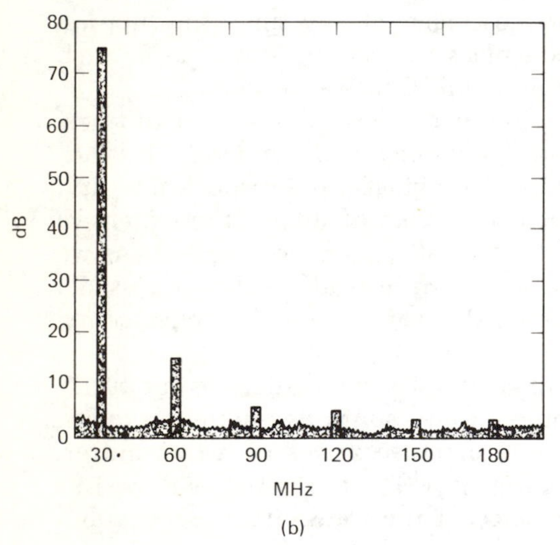

Figure 1.5 (a) Graphical representation of the second- and higher-order harmonics from a transmitter that would not comply with FCC regulations; (b) an acceptal spectral output where all harmonics are 60 dB or greater below peak power. The vertical line at the far left represents the desired output energy.

greater than 40 dB below peak power; hence they meet the FCC rules and regulations.

After having the equipment performance rejected by the engineering supervisor, or through self-decision, the engineer designs a proper harmonic filter. The results might be like those of Fig. 1.5(b). The most significant harmonic (second) is down 60 dB from peak power, and the remaining responses are better than 65 dB down. This kind of performance is not difficult to obtain with the design techniques that are in force today.

The spectrographs of Fig. 1.5 do not show other kinds of spurious responses that are possible with transmitters that use the heterodyne frequency-generation technique. In a real-life situation we would undoubtedly see a number of additional responses that were not related harmonically to the fundamental signal. There will be more on this subject later in the book.

Another matter of concern in terms of the purity of the output energy of a transmitter is *intermodulation distortion* (IMD). The poorer the IMD characteristic of a linear amplifier, the broader the transmitted signal will be. With the technology available to us at present it is an easy matter to design a transmitter that can produce a reasonably clean output in terms of IMD.

An example of the type of display we might see on a spectrum analyzer when conducting a two-tone IMD test is shown in Fig. 1.6(a).

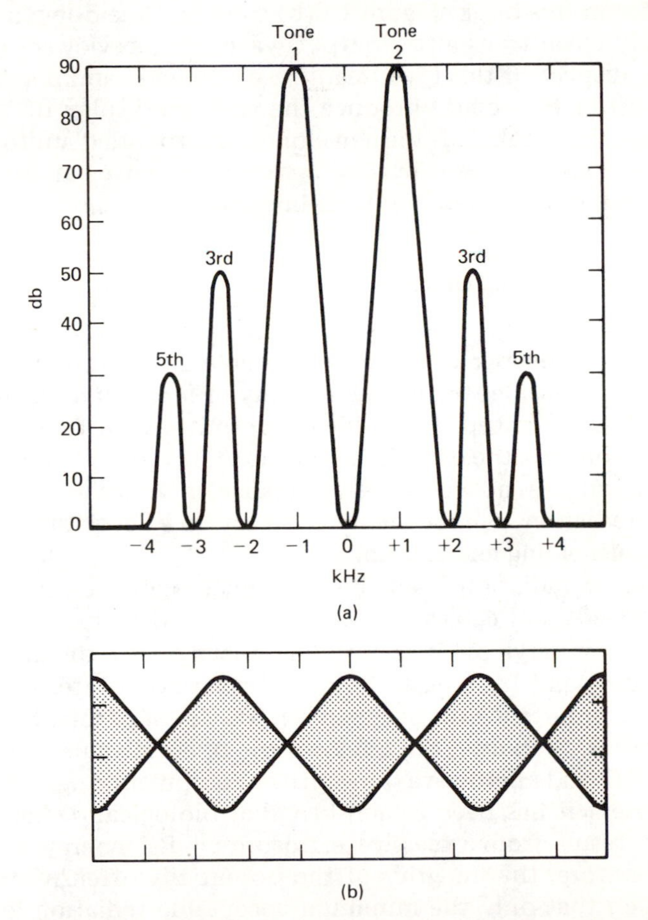

Figure 1.6 (a) Spectral presentation of the distortion products (IMD) from a transmitter to which a two-tone test signal has been applied. The third-order products are 40 dB below peak power, and the fifth-order products are 60 dB below peak power; (b) a time-domain display of a reasonably clean transmitter output, as observed on a scope.

This is called a frequency-domain signal display. It approaches the ideal condition. In a practical situation we would see additional responses, noise, and asymmetry of the third- and fifth-order distortion product responses. This illustration was used merely to characterize the two tones and their third- and fifth-order responses. The higher the third- and fifth-order peaks, the greater the amplifier distortion, and hence the greater the bandwidth of the signal. This spectrograph shows the third-order response at -40 dB, with the fifth-order response down some 60 dB from peak power. Although this is a typical (good) condition for a well-designed transmitter, some commercial equipment exhibits an IMD profile that is much inferior to the example shown. Methods for ensuring minimum distortion in linear amplifiers will be treated later in this book. Figure 1.6(b) shows a time-domain display of a reasonably clean transmitter output waveform, as viewed on an oscilloscope. A display of this type might result when examining the output of a transmitter that could produce the spectral display of Fig. 1.6(a). Notice that the peaks of the envelopes are rounded uniformly, and there is no separation or extension between the cusps at the base line. This is a good indication of amplifier linearity.

1.4.3 Incidental Radiation

As we look deeper into the fundamentals of transmitter design, it is vital that we consider the matter of stray or *incidental radiation*. Continuing with the concept of an ideal transmitter, we find that RF energy existed only at the designated output port (antenna terminal). Unfortunately, this is not possible to realize in a practical design, but incidental radiation can be held to a very low level through careful attention to shielding and filtering.

Incidental radiation from the transmitter cabinet and its external connecting leads and cables can be the cause of harmful interference to other broadcast services. It is for this reason that a designer must include shielding and filtering techniques that prevent interference in the near field to entertainment devices such as television sets, FM receivers, and AM radios. Similarly, the safety of the operator must be ensured in the event UHF and microwave transmitting equipment might be "leaky." In this context it has been established that biological damage to living tissues can result from excessive exposure to RF energy at UHF and higher. Therefore, the integrity of the potentially offensive transmitter must be such that only the minimum acceptable radiation levels exist.

Basically, preventive measures include cabinet bonding where metal surfaces mate, minimum apertures in the cabinetry, and filtering of the related cables (such as power leads, microphone cables, and control lines). Some equipment contains double shielding (one cabinet inside another) to enhance the quality of the shielding. We will learn more about this subject elsewhere in this volume.

1.4.4 Modulation Characteristics

Among the basic transmitter-design characteristics that are worthy of discussion are the modulation traits for a given transmitter. The major objective for any type of voice-operating mode is *minimum distortion* of the transmitted intelligence. As we learned earlier in this chapter, the greater the amount of distortion, the wider is the transmitted signal. Apart from that, it is important to realize that the higher the distortion level, the poorer will be the readability or "comprehension" of the received signal.

Irrespective of the voice mode under consideration, the transmitter should be capable of transmitting only the desired mode. That is, an AM transmitter should not contain FM along with amplitude modulation, or vice versa. Similarly, a single-sideband (SSB) emission should not have AM and FM components included in the output waveform, even though SSB is a special form of AM. In a like manner, visual carriers from TV transmitters should be free from FM rather than a mixture of AM and FM.

Because of the foregoing ground rules, it becomes necessary for the designer to analyze the completed circuit to ensure that there are no unwanted modulation-mode components present in the transmitter output. An excessive level of these parasitic modes can greatly impair the readability of a signal. Excessive FM in an SSB or AM signal will increase the bandwidth.

We must pay attention also to the modulation spectrum of a voice or video transmitter. This is done in the early stages of the equipment by tailoring the speech or video amplifier bandwidth to the service the equipment will be used for. The FCC bandwidth specifications for the particular broadcast service must be taken into account as well.

Once the design is finalized and the transmitter complies with government regulations, we need only to monitor the transmitted signal for excessive modulation percentage or excessive deviation in the case of an FM transmitter. Through the concept of signal monitoring we can also determine whether too little modulation or deviation exists.

Graphic examples of modulation percentages, for an AM transmitter are shown in Fig. 1.7 as they would appear on an oscilloscope. Figure 1.7(a) illustrates *undermodulation* (less than 100%). In Fig. 1.7(b) we can see the effects of *overmodulation* (more than 100%). Modulation in excess of 100% is the primary cause of excessive signal bandwidth and distortion. A 100% modulation characteristic is presented in Fig. 1.7(c). The modulation cusps should be symmetrical in a perfect case.

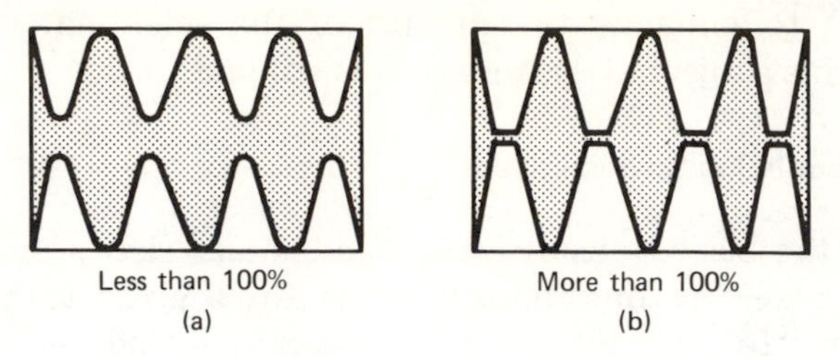

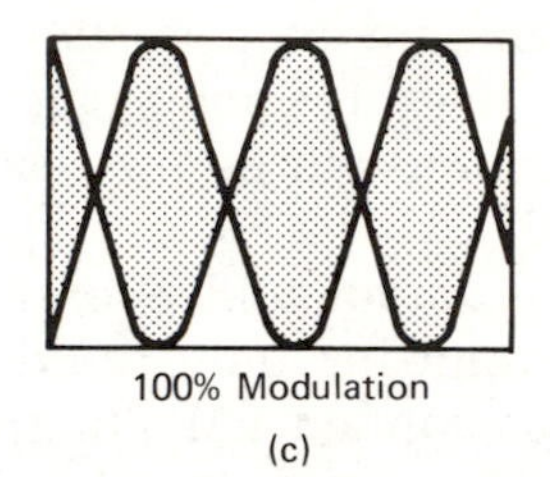

Figure 1.7 Waveforms of an amplitude-modulated transmitter. Waveform (a) shows undermodulation, while waveform (b) represents overmodulation (distortion); (c) a proper waveform for 100% modulation.

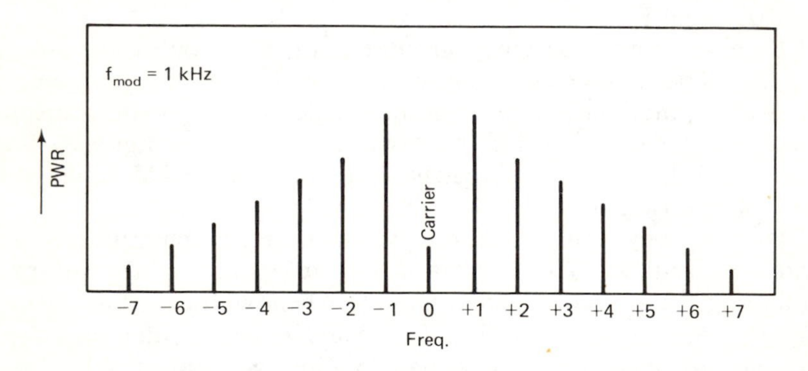

Figure 1.8 Displacement of discrete sidebands from an FM transmitter with a 1-kHz tone applied to the modulator.

Displacement of the discrete but *significant sidebands* of an FM transmitter is shown in Fig. 1.8. A 1-kilohertz (kHz) modulating frequency is specified in this presentation. It can be seen that significant sidebands will appear at 1-kHz intervals above and below the carrier frequency. In a properly adjusted transmitter the sidebands will be symmetrical, as shown. The greater the applied modulation, the greater is the *deviation* or bandwidth of the signal.

Interference can be generated by a poorly designed continuous-wave (CW) transmitter, as is the situation with voice-mode transmitters. Excessive bandwidth will result if the CW waveform lacks proper *shaping*. A worst-case example is seen in Fig. 1.9(a). The leading and trailing

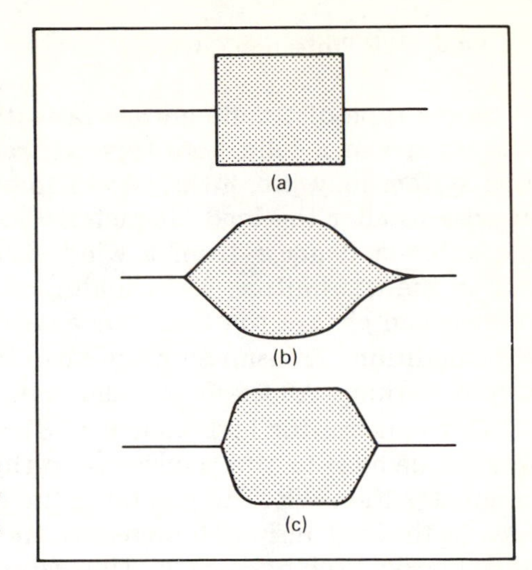

Figure 1.9 Waveforms showing the effects of (a) insufficient CW wave shaping, (b) excessive shaping, and (c) proper shaping.

edges of the RF waveform are abrupt. This will cause severe *key clicks*, which can spread over several megahertz when the transmitter is operated. Excessive shaping is shown in Fig. 1.9(b). The intelligibility of the signal, because of excessive (soft) shaping, can suffer markedly under poor-signal conditions. A waveform with proper shaping of the leading and trailing edges is shown in Fig. 1.9(c).

1.4.5 Overall Efficiency

As designers, we must consider the overall efficiency of our composite transmitter. Apart from the matter of energy conservation, we need to minimize the power consumed by the equipment to conserve in terms of battery drain. Reduced power consumption lessens the amount of heat generated by the active devices in a transmitter. This is important in terms of circuit stability and longevity.

Minimizing battery drain is of special importance with regard to hand-held transceivers, land-modile units, and airborne apparatus. It is incumbent upon the designer, therefore, to study each stage of the design in terms of operating current and efficiency. The decision to use solid-state switching circuits, as opposed to those which contain relays, is part of this important exercise. The efficiency of the active stages must be regarded carefully (and optimized) during the design period. Network losses must be low so that driving power can be kept to a minimum between all the power stages of a transmitter.

Still another matter to ponder is that of using analog frequency readout instead of digital circuitry with its power-consuming frequency counters: Analog readout systems are entirely adequate in many instances.

1.4.6 SWR Protection Circuits

Some type of standing-wave ratio (SWR) shutdown circuit should be contained in a solid-state type of transmitter. Even though the antenna system may present a correct load for the transmitter, problems can arise to alter the load characteristics. For example, we may experience antenna damage from a wind storm, thereby presenting a high SWR to the transmitter. The buildup of ice and snow on some kinds of antennas can change the impedance seen by the transmitter, causing an SWR condition. Transmission lines can become open- or short-circuited without warning. All these problems can cause a high SWR.

Our concern for high values of SWR (generally 2 : 1 or greater) is based on damage to the transistors in the final stage of the transmitter. The greater the SWR, the greater is the mismatch seen by the amplifier stage. As the load mismatch increases, the match between the transistors and networks becomes worse. This causes unusually high peak voltages in a worst-case example, and this can destroy the transistors. Another cause of mismatch danger is the change in network Q as the load on the transmitter departs from the design value (usually 50 Ω). As the collector-network Q rises, there is a strong probability of self-oscillation occurring at or below the operating frequency. If the self-oscillation is vigorous enough, the amplifier transistors can be subjected to excessive peak voltages and heat.

Some form of SWR sensor should be included in the design. It can be used to control an SWR-protection circuit, which lowers the drive to the power amplifier as the reflected component from the antenna increases. A basic circuit of this type is shown in Fig. 1.10. $T1$ is a toroidal transformer that is used to sample the 50-Ω transmitter output. When an SWR of 1 exists, there is no dc voltage developed by $D1$ and $D2$, a voltage-doubler circuit. As the SWR rises from 1 : 1, $D1$ and $D2$ rectify the reflected RF voltage on the line and reverse-bias the control transistor, $Q3$. Since a positive control voltage is developed by the rectifier diodes, and because $Q3$ is a *pnp* device, the positive base-control voltage moves $Q3$ away from full saturation and decreases V_{cc} to the driver stage, $Q1$. $Q3$ is saturated normally by virtue of the 2.2-kilohm (kΩ) base-return resistor. In practice, a more elaborate system than that seen in Fig. 1.10 would be used. $C1$ is adjusted for zero dc output from $D1$ and $D2$ when the SWR is unity, or 1. The transmission-line center conductor passing through the toroid core of $T1$ constitutes a single-turn primary for the transformer. The secondary winding of $T1$ has several turns of wire.

VHF and UHF transmitters generally contain a different style of SWR sensor than that shown in Fig. 1.10. Many designers prefer to use a strip-line style of 50-Ω sensor that contains a pickup line for sampling the reflected power. This type of circuit is seen in Fig. 1.11. Its ad-

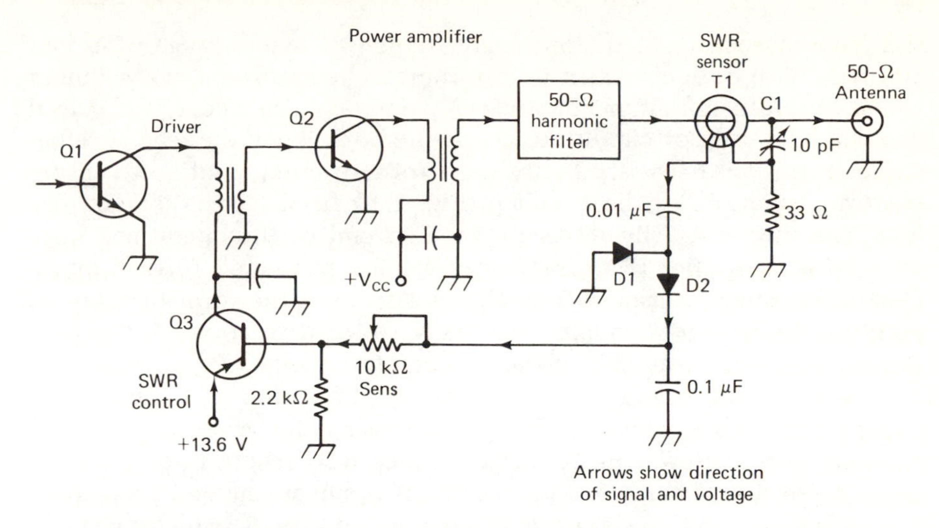

Figure 1.10 Technique for offering VSWR protection to the final amplifier of a solid-state transmitter. A VSWR sensor develops control voltage that reduces the driver output power in the presence of excessive reflected power from the antenna.

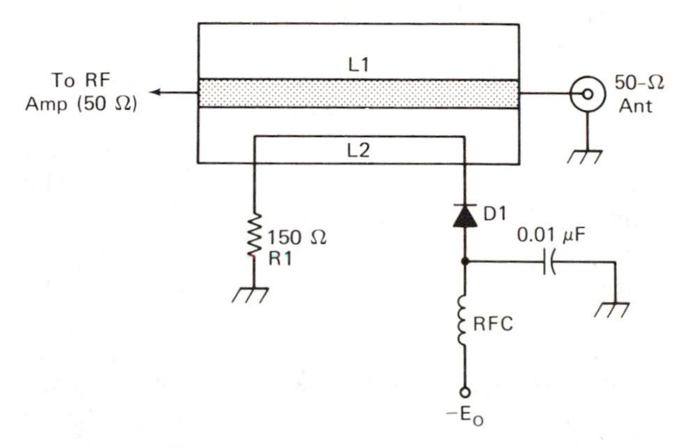

Figure 1.11 Details of a strip-line type of VSWR sensor.

vantage at VHF and UHF is that it presents a flat load to the amplifier because of its 50-Ω strip-line format. The SWR sensor of Fig. 1.10 is hard to null above 30 MHz and could easily present a "bump" in the 50-Ω line, causing an SWR condition *within* the transmitter.

The line sampler in Fig. 1.11 is constructed of printed-circuit board that has a copper ground plane. *L*1 is dimensioned to be a 50-Ω line on the ground plane. *L*2 is a small sampling line that is spaced approximately 3 millimeters (mm) from *L*1. As the reflected power increases and flows through *R*1, the voltage from *D*1 becomes greater. This circuit suffers from *frequency sensitivity* and is useful over only a few megahertz at VHF and UHF. Conversely, the circuit of Fig. 1.10 is relatively insensitive to the operating frequency over a range of 1.5 to 30 MHz. With both circuits it is important to keep unwanted reactances to an absolute minimum. Parasitic X_c and X_L can occur if the lead lengths in the sensor circuit are not kept short. When the reactance becomes high, it is necessary to use some form of compensation to ensure uniform operation of the circuit in Fig. 1.10 from 1.5 to 30 MHz. The compensating is usually done at the high end of the operating range.

Power-amplifier (PA) transistors can be protected further by inclusion of a heat sensor. Often, the sensor takes the form of a silicon junction diode that is coupled thermally to the case of one of the transistors, or to the heat sink in the immediate vicinity of the transistors. As the heat rises to a critical value, the sensor's internal characteristics change (internal resistance). This permits using the sensor as a control element to shut down the PA stage, reduce RF drive to it, or actuate a cooling fan that lowers the temperature of the heat sink and transistors. The sensor is used to trigger a larger control circuit, which normally contains dc amplifiers and/or switches that cause the required control function to occur.

1.4.7 Frequency Readout Resolution

Some transmitters are designed to cover a wide range of frequency in continuous-coverage fashion. When an analog-readout scheme is employed, it is difficult to provide resolution better than 1 kHz. An ideal readout system would exhibit finite resolution, or at least 1-Hz frequency increments. This is an area for which no national or worldwide standards exist. Resolution in excess of 1 kHz (5 kHz, 10 kHz, and higher) is archaic by present-day standards and should probably not be considered in any new design. The 1-kHz increment system seems to be entirely adequate for many fixed-station operations, since calibration checks can be made easily with accessory devices, such as frequency counters.

Most transmitters that use frequency synthesizers, and that are designed for wide-range continuous-coverage operation, use 100-Hz increments with click-stop dial mechanisms. These require digital frequency-readout provisions, since analog mechanisms would be impractical with that level of resolution.

If an *LC* type of oscillator is used in an analog-readout system, it should be of the linear master oscillator (LMO) variety. This will help ensure dial accuracy and linearity when tuning from one end of a frequency range to the other. The greater the linearity, the less frequent is the need to recalibrate the dial or local oscillator (some dials with 1-kHz increments can be slipped for calibration purposes).

The need for linearity is not so great when digital frequency displays are used. As in the case of analog readout methods, there seems to be no standard regarding how far to go in terms of readout resolution. Some designers of HF-band equipment are satisfied with 100-Hz resolution (e.g., 16,110.1 kHz). The resolution is usually in megahertz *and* hertz at VHF and higher. There the designer may opt for three places to the right of the decimal point (e.g., 151.335 MHz). But in either case the last significant figure is in hertz, and the ultimate resolution remains in 100-Hz steps. The frequency can, of course, be carried as far to the right of the decimal as desired, consistent with design practicality.

Most digital readout systems contain light-emitting diodes (LEDs) or liquid-crystal display (LCD) devices. The choice depends partly on how much cognition is possible (in terms of room lighting) in the environment where the equipment is used. In a broad sense, the LED display will yield the best results in darker rooms, whereas the LCD display may be superior in bright areas.

1.5 RECEIVER PERFORMANCE

The criteria for satisfactory circuit performance in communications receivers is not significantly different from that which was outlined for transmitters. However, there are some performance considerations we must honor that do not pertain to transmitters. These will be used as our focal point in this section. Redundancy will be avoided by not discussing frequency stability, incidental radiation, overall efficiency, and tuning and readout resolution. These subjects apply equally to transmitters and receivers, as do the design objectives we have already discussed.

1.5.1 Dynamic Range

Assuming that all design objectives are met in a particular receiver design, the single most important performance characteristic is *dynamic range*. A receiver with otherwise excellent design features can be a total loss to a user if it cannot deliver high performance in the presence of strong signals. Dynamic range is the difference, in decibels, between the overload point and the *minimum discernible signal* (MDS) or acceptable signal level in a receiver. The MDS is established by the internal noise level of the receiver. Generally, the MDS point at which measurements commence is 3 dB of signal above the noise, as referenced to the receiver noise floor. This is considered to be a signal level that could be detected by the receiver. This measurement is made at the receiver output by means of an audio voltmeter.

The significant receiver performance parameters for determining the dynamic range are *blocking* and *intermodulation distortion* (IMD) levels. Both are referenced to the MDS level just discussed. Blocking is observed as *gain compression*. Two signal generators are used for this evaluation. One is set for a low output level, while the other is adjusted for a frequency that is 20 kHz away from that of the first generator. The signals are fed into the receiver under test through a hybrid combiner, thereby preventing one generator from phase or frequency modulating the other. When the generators have been set in the foregoing manner, the output of generator 2 is increased until there is 1 dB of decrease in the level of the weak signal from generator 1, as noted on the audio voltmeter at the receiver output. This measurement indicates the maximum input signal level the receiver can accommodate before desensitization or compression occurs. A block diagram of a typical measurement setup for receiver dynamic range is presented in Fig. 1.12.

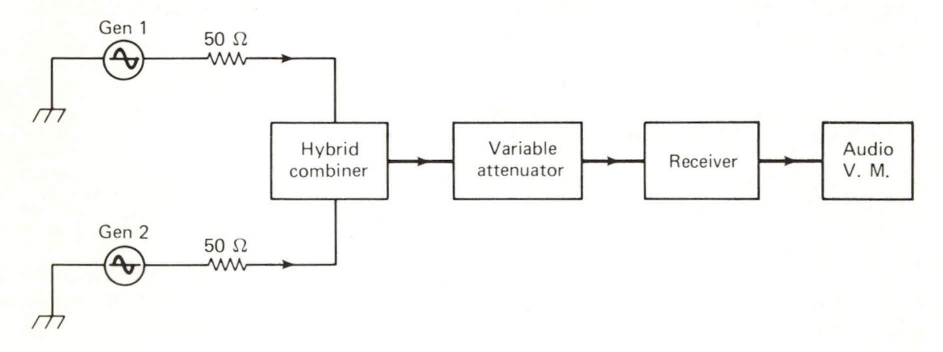

Figure 1.12 Block diagram of a test setup for measuring receiver dynamic range.

The two-tone IMD characteristic of a receiver is probably the most significant of the receiver dynamic-range traits. These distortion products can be developed in the RF amplifier, mixer, and IF system (including the IF selectivity filters). Generally, the mixer is the worst offender we will encounter with regard to front-end IMD. But there can be situations where the RF amplifier and mixer yield excellent IMD traits, while the IF filter (because of poor mechanical construction and inherent limitations) establishes the actual IMD profile. We can see from this that the IMD performance of a receiver is a collective characteristic, and the filter or active and passive stages of the receiver can singly degrade the IMD even though the remaining circuit elements exhibit excellent performance.

The IMD measurement is made while using two signal generators with their outputs set 20 kHz apart (Fig. 1.12). The signals are set for approximately -10 dBm and fed into the receiver through a hybrid combiner and a step attenuator. We will refer to the two frequencies as $f1$ and $f2$. As we proceed with our test, we will observe some third-order IMD products. They will be apparent in the receiver at $2f1$ minus $f2$ and at $2f2$ minus $f1$. Hence, if the generators are set for 10.050 and

10.070 MHz, we will note the third-order IMD responses at 10.030 and 10.090 MHz.

In determining the IMD characteristic, we must tune the receiver to one of the third-order frequencies, say, 10.090 MHz. Next we will adjust the step attenuator until the third-order product rises 3 dB above the MDS level (noise floor), as observed on our audio voltmeter.

Once the preceding condition is met, we can proceed with our determination of the dynamic range. For example, if the attenuator ended up with 40 dB of attenuation to the two -10-dBm generator signals, and the hybrid coupler provided its normal 6 dB of loss, the IMD level would be -10 dBm, minus 6 dB, minus 40 dB. This being the case, our IMD level would be -56 dBm.

If we reference this to the receiver MDS level, our IMD characteristic becomes MDS minus IMD. Assuming we had a hypothetical MDS of -135 dBm, our dynamic range (DR) would become -135 dBm minus -56 dBm, or 79 dB. This would imply a fairly good receiver, but not an excellent one. Dynamic-range levels in excess of 100 dB are possible with the present state of the art. Blocking levels considerably in excess of that amount are common. The measuring technique just described is used by many receiver designers and analysts. The method was suggested to the author several years ago by W. Hayward of Tektronix Corporation. It has proved to be entirely viable for receiver-performance evaluation.

1.5.2 Effects of Poor Dynamic Range

We might wonder how a limited dynamic range would affect the actual performance of a communications receiver. First, strong in-band and out-of-band signals could cause receiver desensitization (blocking) so severe that the desired signal could not be copied. The weaker the desired signal, the more pronounced would be the effect.

Poor dynamic range can lead to cross-modulation of received signals. That is, an adjacent signal in the band of interest can become superimposed on the desired signal to render the readability impossible.

Poor IMD performance can cause spurious signal responses across the tuning range of a receiver. These responses may appear as additional strong signals, thereby causing unnecessary interference. Responses of this variety can appear on the frequency of interest to ruin the intelligibility of a signal.

Later we will learn methods of design that will ensure high dynamic range through proper gain distribution and selection of the devices used in the receiver front end. It was important in this chapter to define dynamic range and discuss its effects, as this is a vital consideration when designing a receiver.

1.5.3 Sensitivity

As receiver designers we must pay attention to the parameter called *sensitivity*. Receiver sensitivity is essentially that quality which permits reception of very weak signals. Contrary to a popular misconception, sensitivity is not defined as a large amount of front-end or overall receiver gain. Rather, it deals with the *noise figure* (NF) of the receiver.

Although excessive internally generated receiver noise can become manifest anywhere in the receiver amplifying stages, the most common source of noise is the receiver front end (RF amplifier, mixer, and post-mixer preamplifier). In essence, noise figure or sensitivity is a measure of the receiver's ability to respond to weak signals without the internal noise masking those signals. External noise does not figure into this parameter (noise arriving via the antenna and the feed line).

One method of measuring the receiver noise figure is through the use of a noise generator and an audio-power meter or voltmeter. A block diagram of this measurement scheme is shown in Fig. 1.13. The

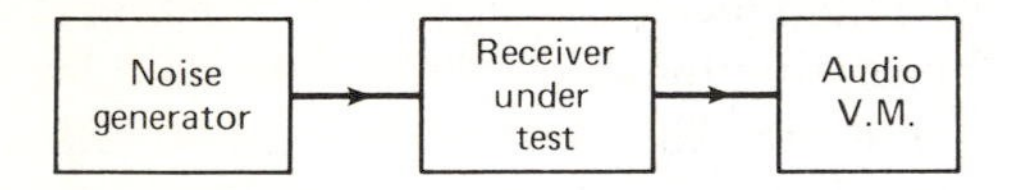

Figure 1.13 A technique for measuring receiver noise figure.

impedance of the noise generator must be the same as the receiver input impedance for the test to be valid. First, the receiver audio gain is advanced until a convenient reference level is established on the audio meter. For illustrative purposes, let's assume that the reading is 6 dB. Next the noise generator output is increased to some value above the 6-dB reference. In this test we will advance the noise output until a 20-dB meter reading is noted. These two noise levels, respectively, will be thought of as $N1$ and $N2$. The noise figure can now be obtained from

$$\text{NF} = \text{excess dB} - 10 \log \left(\frac{N2}{N1} - 1 \right)$$

Therefore, our receiver noise figure is

$$\text{NF} = 14 - 3.67 = 10.3 \text{ dB}$$

This would be an acceptable receiver noise figure for reception below 30 MHz, since the antenna noise (atmospheric and man-made) generally exceeds this level of receiver noise. Perhaps a more ideal noise figure for an HF-spectrum receiver would be on the order of 5 dB. At VHF and higher, the noise figure becomes more significant because the antenna noise is very low, comparatively. Noise figures of 1 dB or less are usually the objective above 30 MHz. Other methods of noise-figure measurement are employed in the industry. A popular measurement technique rates the noise characteristics of the receiver with respect to signal input. The resultant expression is $S + N/N$, as given in decibels.

As designers, it is our objective to strive for a noise figure that is suitable for the frequency of receiver operation. This requires a careful choice of active devices in the front end of the receiver. Not only should these devices be of the low-noise variety, they must also be capable of ensuring high dynamic range. The operating voltages and signal injection levels need to be set for the lowest noise figure obtainable from the devices used. We will learn more about this later.

1.5.4 Selectivity

Perhaps the second most important performance characteristic we must acknowledge as receiver designers is *selectivity*. If the receiver has selectivity, it will be capable of differentiating between the desired signal and those signals that appear inside and outside the band of interest.

Selectivity can be introduced at various points in a receiver circuit, as shown in the block diagram of Fig. 1.14. At each of the locations indicated by $FL1$, $FL2$, $FL3$, $FL4$, and $FL5$, a specific advantage is realized. First, $FL1$ and $FL2$ constitute what is commonly referred to as *front-end selectivity*. The front-end filter or filters are used to restrict the passage of signals above and below the frequency of interest. This aids receiver dynamic range and protects the mixer stage from strong unwanted signals that could cause *image frequencies* (unwanted signal responses or mirror images of signals that are related numerically to the IF or local-oscillator frequency). The objective is to reduce the image response as much as possible through the employment of selective tuned circuits or filters between the antenna and the mixer.

Examples of how we might accomplish the goal of selectivity are given in Fig. 1.15. Fig. 1.15(a) and (b) show tunable resonators being used as filters. The higher the loaded Q (Q_L) of the resonators ($C1A/L2$ and $C1B/L3$), the greater is the selectivity. Q_L becomes higher as the coupling to the load and antenna is reduced. This can be controlled by varying the turns of wire at $L1$ and $L4$ or by the physical placement of $L1$ and $L4$ with respect to $L2$ and $L3$. Top-coupling capacitor C_c is chosen to prevent a dip or slump in the response curve of the two resonators. Too large a capacitor value at C_c will lead to a double-hump response, rather than the desired single-hump response. All filters must have some loss in order to function as filters. In circuits like those of Fig. 1.15(a) and (b), it is not unusual to adjust for an insertion loss (IL) as great as 6 dB by using very light input and output coupling. The higher the IL the better the Q_L, and hence the higher the selectivity. The IL can be recovered by means of $Q1$.

By increasing the number of resonators in a tunable filter, we can increase the skirt selectivity $FL1$ of Fig. 1.15(b). A three-section vari-

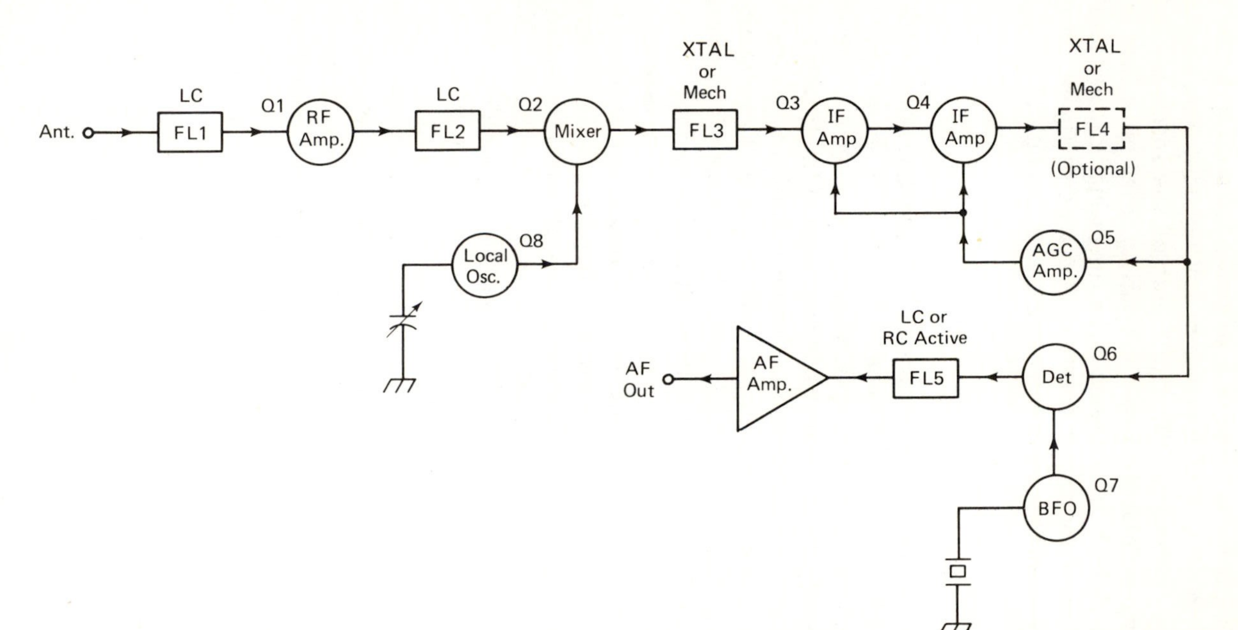

Figure 1.14 Block diagram of a typical single-conversion superheterodyne receiver. IF and audio filtering is indicated.

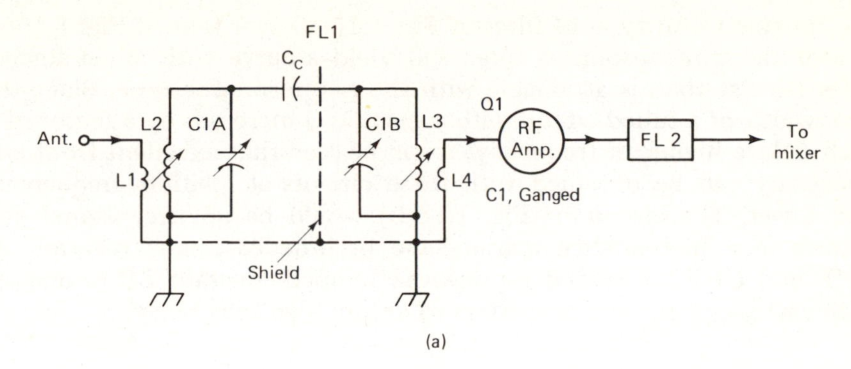

(a)

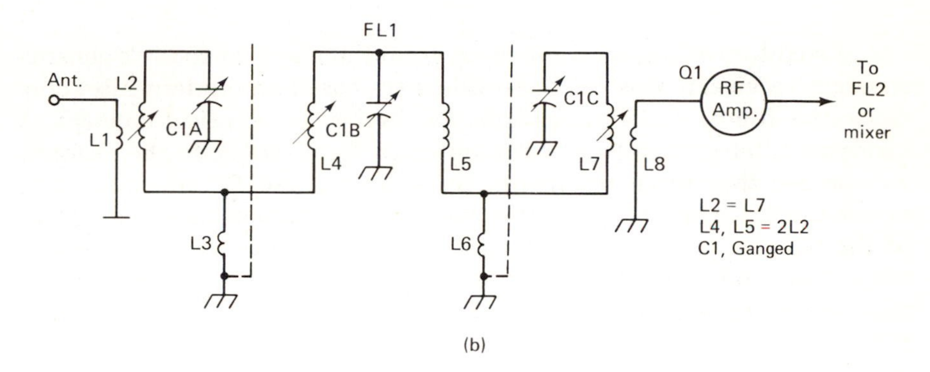

(b)

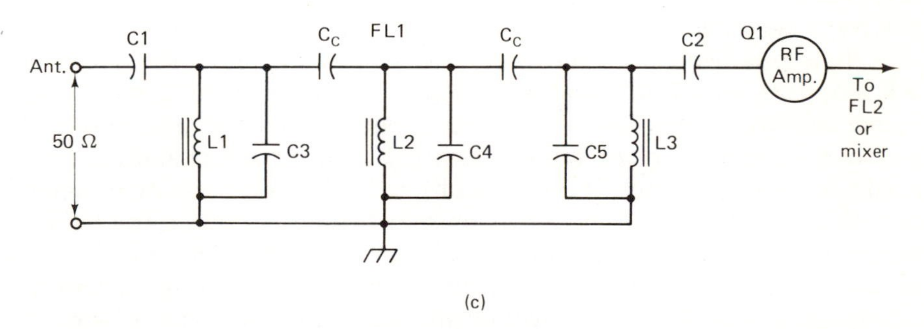

(c)

Figure 1.15 Various techniques for ensuring good front-end selectivity in a receiver: (a) and (b) tunable networks; (c) a broadband filter.

able capacitor can be used to tune three resonators, as seen in Fig. 1.15(b). Since $L4$ and $L5$ are in parallel and are tuned by $C1B$, they require twice the inductance used at $L2$ or $L7$. In this circuit we can see the application of *bottom coupling* by means of small inductors ($L3$ and $L6$). They are adjusted in the same manner as C_c of Fig. 1.15(a) to minimize the manifestation of a double-hump response. Loading of the filter is done via $L1$ and $L8$, as discussed earlier. The advantage of the tunable Cohn type of filter of Fig. 1.15(b) over that of Fig. 1.15(a) is that the three-resonator filter will yield a curve with much steeper sides (skirts) than is attainable with the two-resonator type. Since the bandwidth of a tuned circuit with a specific Q increases by a factor of 2 each octave higher in frequency, it can be seen that excellent front-end selectivity can be obtained with these circuits at medium frequencies and lower. The circuit of Fig. 1.15(b) would be an exceptional performer in a high-quality marine-band or broadcast-band receiver. At VHF and UHF we would replace the lumped-constant LC resonators with cavity or strip-line resonators to obtain high selectivity.

1.5.5 Fixed-Tuned Filters

Fixed-tuned front-end filters are found in many of today's general-coverage receivers. Most of these filters are based on modern filter concepts and follow the Butterworth, Chebyshev or elliptical concept. A bandpass filter of this type is illustrated in Fig. 1.15(c). In the example we can see that three resonators are used. $C1$ and $C2$ are selected to match the input and output impedances to the characteristic impedance of the filter. C_c serves as the coupling capacitor between the filter sections. The bandwidth (BW_L) of this kind of filter is substantially greater than that obtained from the circuits of Fig. 1.15(a) and (b). A typical BW_L at 10 MHz might be 500 kHz. Therefore, $FL1$ cannot reject adjacent signals within the filter passband, but it will attenuate signals above and below the design cutoff frequencies (half-power or 3-dB points on the response curve).

It is practical in all the circuits shown in Fig. 1.15 to follow the RF amplifier with a duplication of $FL1$. We may decide that the additional selectivity provided by a second front-end filter is unnecessary. If so, then a simple tuned circuit, consisting of a capacitor and inductor, could be used to match the RF amplifier to the mixer. Some designers use untuned broadband inductors or transformers between the RF amplifier and the mixer. Others place $FL1$ between the RF amplifier and mixer, while using a simple tuned circuit between the antenna and RF amplifier. A classical design would probably contain a highly selective filter ahead of and after the RF amplifier. Cost, assembly time, and the intended use for the receiver are usually the determining factors in such a decision.

Figure 1.14 shows $FL3$ immediately after the receiver mixer. In some designs we will find a postmixer IF amplifier ahead of the IF filter. In either event, $FL3$ establishes the *IF selectivity* profile. This filter can be a mechanical-resonator type, a crystal-lattice filter, or a piezo-crystal unit. The latter is often called a *monolithic filter* because it is formed as a single crystal element. A crystal-lattice filter contains two or more discrete crystal elements. In all instances, the center frequency of the filter (center of the passband) is the desired intermediate frequency of the receiver. Thus, if the chosen IF were 10.7 MHz, the center of the filter passband would have to be 10.7 MHz also.

The bandwidth of the IF filter is chosen for the mode of reception desired. An AM filter, for example, might have a 10-kHz bandwidth at the 3-dB points of the curve. Narrower AM filters are used when fidelity is not a primary concern to the user. For single-sideband (SSB) reception, we would use a 2.2-, 2.4-, or 2.7-kHz filter bandwidth, depending on how much fidelity-cutting we could tolerate. For reception of CW and teletype (RTTY) we would choose a very narrow filter. For CW use the bandwidth may be as low as 200 Hz or as great as 1 kHz, depending on the service to which the receiver is put. But the narrower the filter response, the better is the rejection of interfering signals that are near in frequency to the signal of interest.

Figure 1.16 shows the circuit for a simple half-lattice type of IF filter. *Y*1 and *Y*2 establish the IF selectivity. They are ground or etched for a frequency pairing that provides the desired bandwidth at *FL*3. R_L is selected to minimize the slump (ripple) between the two crystal frequencies. This is normally a low value of resistance between 250 and 10 kΩ. The closer in frequency the crystals are to one another, the narrower is the selectivity response. The skirts of the response curve will be relatively broad with simple filters of this kind. This is called the *shape factor* (ratio of the bandwidth at the 6- and 60-dB points of the response curve). Therefore, if the -6-dB point on the filter curve was 5 kHz away from center frequency, and the -60-dB point was 12 kHz away from center frequency, the shape factor of the filter would be 12/5 = 3 : 1. An ideal filter would have a shape factor of 1 : 1. This would result from having a bandpass response that was perfectly rectangular. Typically, however, the shape factors of IF filters in communications receivers run from 2 : 1 to 5 : 1.

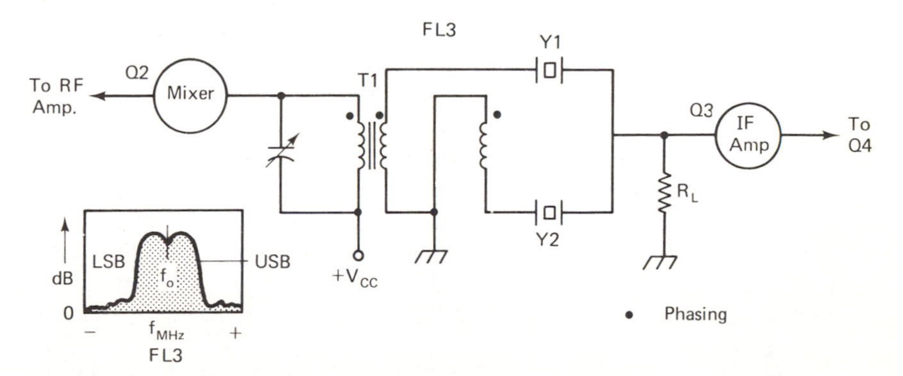

Figure 1.16 Two quartz crystals can be used to form a simple half-lattice IF filter.

As is true of *LC* types of filters (Fig. 1.15), the more sections or poles in the filter, the better is the shape factor. Hence, a properly designed crystal-lattice filter with four poles would have much steeper sides on its curve than a two-crystal filter of the type seen in Fig. 1.16. For SSB reception the beat-frequency oscillator (BFO) output is placed approximately 20 dB down on either side of the IF filter response curve, as illustrated in the inset drawing of Fig. 1.16. The depression between the humps is the filter ripple. Complex filters have numerous small depressions across the "nose" of the filter response, which more correctly qualifies the term *ripple*. For reception of CW signals it is common practice to place the BFO approximately 700 Hz away from the center frequency of the filter, which is the receiver IF. The actual relationship between the IF and the BFO is arbitrary, since some designers prefer offsets as great as 1000 Hz for CW work. The BFO frequency is more critical for SSB reception: if it is not in the proper relationship to the IF, the voice reproduction sounds unnatural (too brassy or high pitched).

IF filters can be used in two places if we follow the plan given in Fig. 1.14. *FL*4 is shown as an optional filter in the block diagram. The advantage of using the second IF filter (tail-end filter) is to reduce wideband noise that originates after the first IF filter, *FL*3. This gives the receiver an improved overall signal-to-noise ratio, a significant advantage during the reception of weak signals. If it is used, *FL*4 should have the same or slightly greater bandwidth than *FL*3. Also, the center frequency of both filters should be the same.

1.5.6 AGC Considerations

Automatic gain control (AGC) has also been called automatic volume control (AVC). In either case the gain-controlling circuitry is important to the comfortable operation of a communications receiver. Without AGC a loud signal can appear in the receiver passband and present an ear-shattering signal in the headphones or speaker if the manual audio-gain control is set for a weak signal. For the most part, AGC is used to level the incoming signal response over a wide range of signal amplitudes. If the AGC circuit provides a high dynamic range, a small signal will have the same loudness as a strong signal; for example, a 1-microvolt (μV) signal at the receiver antenna terminal will appear to be the same strength as a 5000-μV input signal as heard at the receiver output. The major difference in the two signals will be noise. The weaker signal will be accompanied by some atmospheric and/or man-made noise (especially at frequencies below 30 MHz), whereas the strong signal will mask the noise.

AGC is derived from IF or audio energy in the receiver. The choice between audio-derived and IF derived AGC is the designer's, but the most common type of AGC is obtained by amplifying the IF signal after sampling it at the last IF amplifier stage, rectifying it, and then applying it to the RF and IF stages as control bias. Some receiver designers prefer to apply AGC to only the IF amplifiers, as AGC can cause nonlinear operation and degraded noise figure when applied to the RF amplifier of a receiver. Most modern receivers employ AGC dynamic ranges which run between 60 and 120 dB.

Figure 1.17(a) shows a basic AGC circuit of the IF or RF-derived variety. A similar system for audio-derived AGC is seen in Fig. 1.17(b). The circuit of Fig. 1.17 (a) shows that light sampling of the IF signal is accomplished by *C*1. A JFET is used at *Q*1 to minimize loading on the output of *U*2. *D*1 changes the IF energy to direct current for application to the difference amplifier, *U*3. *C*2 and *R*2 are used to establish the AGC time constant. A typical decay time for the AGC in a communications receiver is 0.5 or 1 second(s). The output direct current from *U*3 is set to swing the desired range for controlling the IF amplifier gain. The prescribed voltage range for RCA CA3028A ICs is +2 to +9 V. *R*1 is used to establish the signal level at which we want the AGC action to commence.

The audio-derived AGC example of Fig. 1.17(b) is similar in many ways to that seen in Fig. 1.17(a). In this circuit we are sampling audio voltage rather than IF voltage. The audio is picked off ahead of the AF-gain control so that the sampling level will not be affected by the control pot. Actual AGC circuits will be offered later in the book.

1.5.7 Noise Limiting

Many communications receivers are equipped with circuits that are effective in *clipping*, *limiting*, or *blanking* pulse types of noise. These circuits can operate at RF or audio, and range from simple series- or shunt-diode types to elaborate systems with timing circuits and shaping networks. Circuits for removing or minimizing noise are important during reception of weak signals, especially in vehicular applications where ignition pulse noise is prevalent. Most noise-reduction circuits result in a performance trade-off of some kind. RF types of blankers often cause degradation of the receiver dynamic range, whereas audio clippers and limiters impair the receiver audio quality. The exception to the foregoing is with regard to FM receivers, where the IF limiter stage does not compromise receiver performance.

1.5.8 Audio Circuit Performance

We do not want high-fidelity audio response in a voice or CW type of communications receiver. Rather, our objective should be to limit the audio bandwidth to that range necessary for the chosen mode of

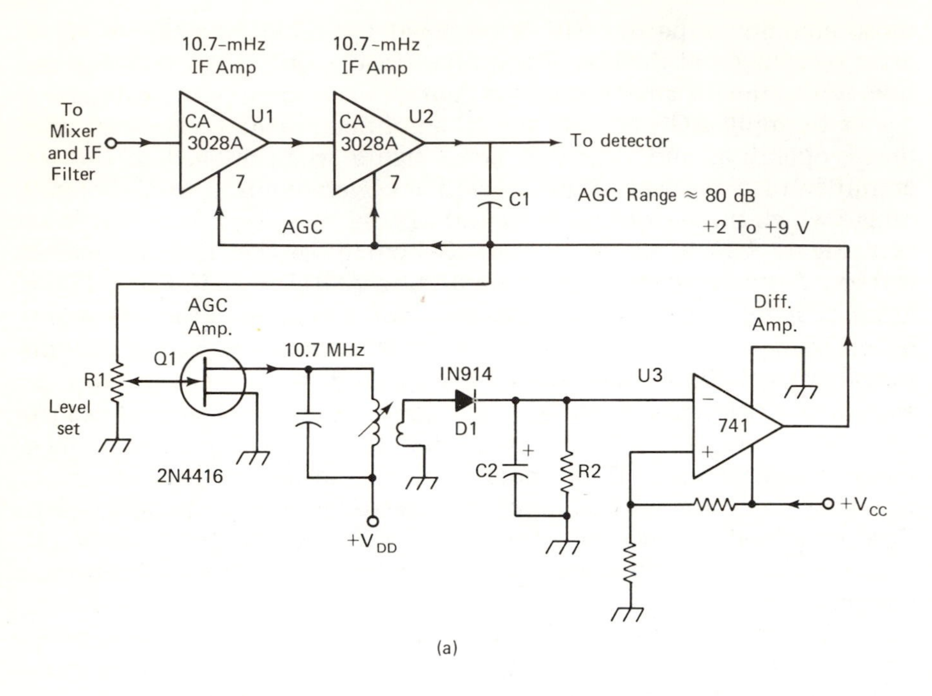

(a)

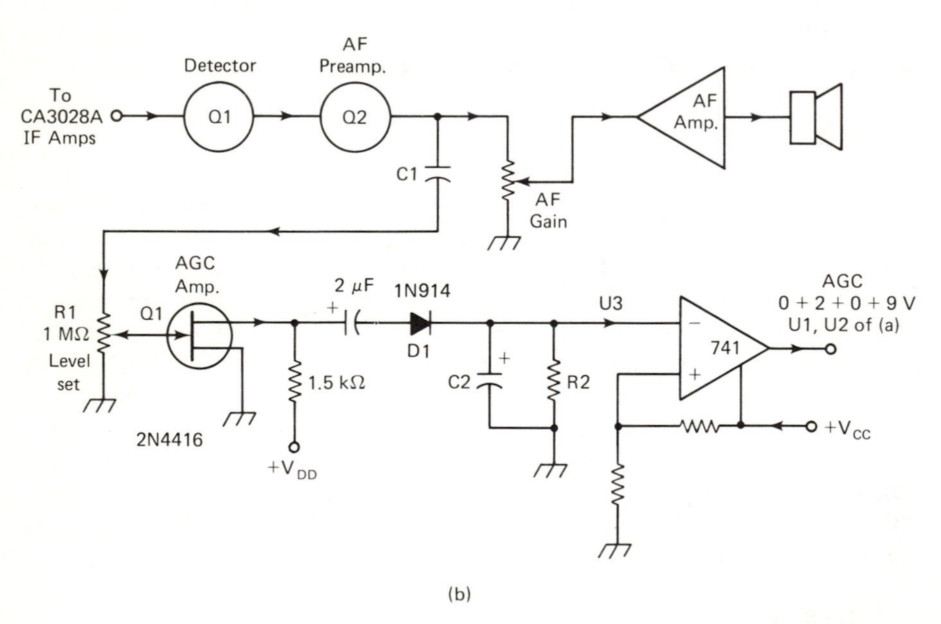

(b)

Figure 1.17 (a) RF-derived AGC, and (b) audio-derived AGC. See Chapter 7 for detailed information on AGC systems.

reception. We can tailor the audio response to our needs by using selected values of resistance and capacitance in the audio-amplifier section or through the addition of audio filters.

Generally, the audio frequencies should be rolled off at approximately 300 Hz on the low side and 2500 Hz on the high side for voice communications. The roll-off points for CW work may be on the order of 200 Hz and 1000 Hz, since 600 to 700 Hz is the CW-note preference of many operators. The advantage in these narrow audio bandwidths is attenuation of low- and high-frequency interference, such as hum, some kinds of noise, adjacent-signal rumble, and high-pitched heterodynes.

Figure 1.18 contains two kinds of audio filters. Figure 1.18(a) shows a passive *LC* filter that yields a bandpass response of the type seen to the right. The more filter sections used, the sharper are the sides

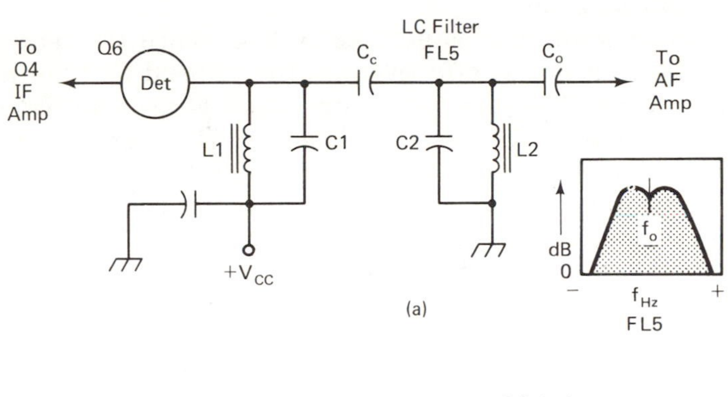

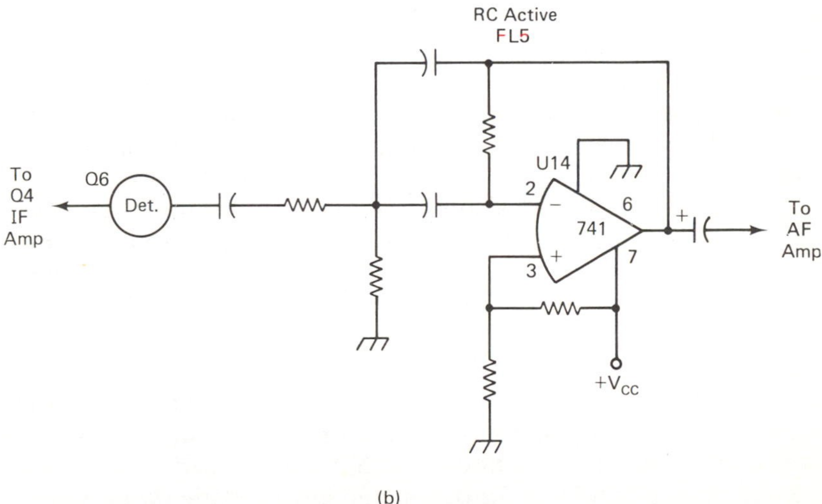

Figure 1.18 Passive or active audio filtering can be used to reduce wide-band receiver noise and provide selectivity for voice or CW reception: (a) passive *LC* filter that will exhibit a small insertion loss; (b) active filter that can provide a gain.

of the response curve. C_c is chosen to minimize the passband ripple (depression). High-Q toroidal or pot-core inductors are employed at $L1$ and $L2$, with high-Q capacitors of good stability used at $C1$ and $C2$. The main disadvantages of LC filters are their large size and some insertion loss.

RC active audio filters are of more value to us because they can be designed to provide gain and very high Q factors. Also, they are physically smaller than LC types of filters. The circuit of Fig. 1.18(b) has only one filter section or pole. Generally, the best performance comes from active filters that contain two to four poles. This allows us to provide steep-sided filter responses.

The LC and active filters of Fig. 1.18 can be configured to provide a high-pass, low-pass, or bandpass response. The bandpass characteristic is preferred by most designers.

A well-designed multipole audio filter can enhance the receiver signal-to-noise ratio, but not as effectively as the tail-end IF filter discussed in Section 1.5.4. The principle of operation is again to restrict the wide-band noise that originates after the first IF filter in the receiver. If the low-frequency cutoff frequency of the filter is, say, 300 Hz, there will be significant attenuation of 60- and 120-Hz hum, which reaches the receiver ahead of the audio filter. Practical circuits for audio filters are presented elsewhere in this volume.

In some respects we can regard the audio channel in a high-performance communications receiver as we would a hi-fi amplifier. Ignoring the wide frequency response of a hi-fi amplifier, let's consider the matter of distortion. If the audio amplifiers in the receiver are not clean (low distortion), readability of weak signals can be impaired markedly. Therefore, we must ensure that our audio circuits operate linearly at the highest output level necessary for the receiver application. Hum and noise should be down 60 dB or more below the weakest signal that can be copied.

A marginal design, based on economic and assembly-time factors, might contain an audio-output IC that could produce only 0.5 or 1 W of reasonably undistorted audio. This might be adequate for some environments (quiet rooms) or when the operator wears earphones. But in noisy rooms we might require 2 or even 5 W of audio to ensure full signal comprehension. A marginal design would be useless in such a situation. Therefore, it is wiser to design for substantially more audio output than is needed, as is done with the better hi-fi amplifiers. This will permit the audio system to operate in the most linear part of its characteristic curve, thus minimizing distortion and ensuring plenty of audio output when it is needed. The power-handling capability of the loudspeaker must be compatible with the amount of audio output expected in a practical situation.

1.5.9 Tuning and Frequency Readout

The recommendations given in this chapter concerning transmitter tuning resolution and frequency readout are applicable to receivers as well. Therefore, we will not cover that subject in this section.

BIBLIOGRAPHY

ARRL, Inc., *The Radio Amateur's Handbook* (all chapters), ARRL, Inc., Newington, Conn. Complete overview and design data for RF communications, plus numerous practical circuits for analog and digital equipment.

DeMaw, D., *Practical Communications Data for Engineers and Technicians*, Howard W. Sams & Co., Inc., Indianapolis, Ind.

DeMaw D., and W. Hayward, *Solid State Design for the Radio Amateur*, ARRL, Inc., Newington, Conn. Design and application of numerous RF communications equipment for various modes.

Orr, W., ed., *Radio Handbook*, Howard W. Sams & Co., Inc., Indianapolis, Ind. (same coverage and scope as the first reference).

2

FREQUENCY-CONTROL SYSTEMS

In the beginning of radio there was one basic form of frequency control for transmitters and receivers, the LC (inductance-capacitance) oscillator. The post spark-gap era transmitter employed the once standard "tuned-not-tuned" or TNT oscillator as a means of frequency control. This was a crude, but then effective, technique that used two triode tubes in push-pull. The grid tank was broadly resonant in the operating range, while the plate tank was made tunable. Many CW transmitters in those days used only a TNT oscillator, which was connected directly to the antenna feeder through a suitable matching network. The efficiency was very poor, owing to a need for light coupling to the load, which minimized load changes on the oscillator, a major cause of oscillator "pulling" and frequency shift. Long- and short-term frequency drift was a serious problem with power oscillators of that variety: this required the use of receivers with considerable bandwidth to keep the transmitted signal in the receiver passband.

Advancing technology brought crystal-controlled oscillators a few decades ago, and stability became the order of the day. Crystal-controlled oscillators remain popular in many communications circuits, especially those that are necessarily compact and power-consumption efficient. The more elaborate piece of equipment might contain a *frequency synthesizer*, but at the expense of greater equipment size and increased power dissipation. The frequency synthesizer is a subject that by itself would need to be treated in a separate volume. Therefore, this book will emphasize crystal and *LC* types of oscillators for use in transmitters and receivers. Modern communications equipment often contains several oscillators. It is not unusual to find a synthesizer being used in the same transmitter or receiver along with several crystal oscillators and at least one *LC* type of variable-frequency oscillator (VFO). This is especially true of equipment that employs the heterodyne frequency-generation scheme.

In this chapter we will discuss basic designs for crystal and *LC* oscillators. All the circuit examples given are practical designs that have been used many times by the author. The design data given are primarily approximations that will yield good performance if quality components are used.

2.1 QUARTZ CRYSTAL

Specific crystalline substances have the ability to develop an electric charge on their surfaces when subjected to mechanical stress. Similarly, these devices will undergo mechanical strain if their surfaces are charged electrically. This phenomenon is known as the *piezoelectric effect*. For use in communications equipment frequency-control and intermediate-frequency (IF) circuits, we will assume quartz as the substance that provides our piezoelectric effect.

A crystal of quartz is formed as a hexagonal prism with a hexagonal pyramid at each end. A typical crystal might be 1 or 2 inches in diameter and several inches in length. Numerous slices or plates can be cut from this crystal to form electromechanical devices used for frequency-control circuits or filters.

If a slab of this quartz is made to vibrate by mechanical means, it will develop an ac voltage between its opposite faces. Conversely, if an ac voltage is applied to the plates of the crystal, it will vibrate at the frequency of the ac voltage. If this voltage is at the same frequency as the natural resonance of the crystal, the vibration of the quartz will be of considerable magnitude. The resultant potential under this condition can reach several volts. If the vibration amplitude is too great, the crystal can become permanently damaged or even fractured.

Slices of quartz are cut from the original crystal at various angles with respect to the *X*, *Y*, and *Z* axes. Each type of cut has a different operational characteristic. The resultant nomenclature for the three most common crystal cuts is *X cut*, *AT cut*, and *BT cut*. In their normal modes of vibration the primary resonant frequency will be dependent upon the thickness of the slab of crystal. The precise relationship depends on the cut angle. As a general rule this relationship is as follows:

Cut Style	f(kHz) × Thickness (mm)
X	2860
AT	1675
BT	1500

To ensure proper oscillation, the crystal plate must be as nearly uniform in overall thickness as possible. The crystal is normally placed between two conductive plates. Some crystals have gold or silver deposited on the flat sides to serve as the electrodes.

A crystal behaves as a series-tuned LC circuit of very high Q. The electrical equivalent of a crystal in its holder is presented in Fig. 2.1.

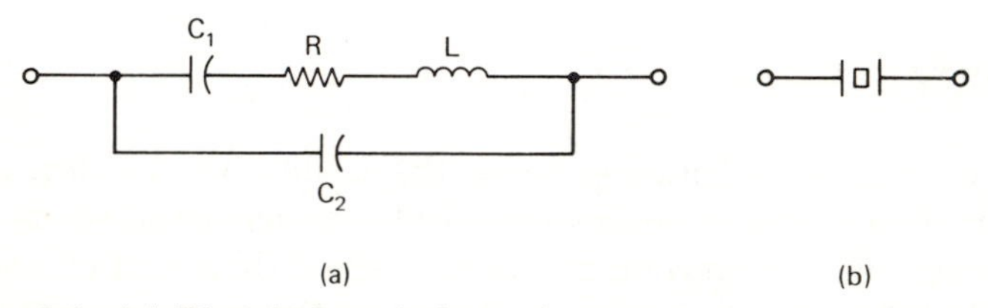

Figure 2.1 (a) Electrical equivalent of a quartz crystal, with $C2$ representing the holder capacitance; (b) symbol for a crystal.

This type of circuit has two frequencies of resonance. One corresponds to a high impedance, while the other is represented by a low impedance. The two responses are shown in Fig. 2.2. The resonant frequency (series) is dependent upon the series combination of $C1$, and L of Fig. 2.1(a). This is the low-impedance condition for the crystal. The parallel, or antiresonant operating mode is brought about by the $C1$-L circuit in combination with $C2$, the parallel capacitance. The difference in operating frequency for the two modes is minor, on the order of a fraction of 1 percent. Most oscillators operate in the parallel-resonant crystal mode. The impedance variation of the crystal is sharply defined for either mode: the impedance can vary by a ratio of 10,000 : 1 or greater when the frequency is shifted only a few hundred hertz. This phenomenon makes quartz crystals especially useful in filter circuits.

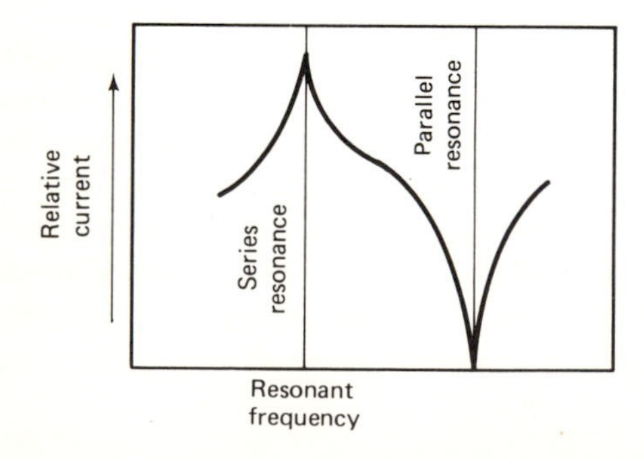

Figure 2.2 Parallel and series resonances occur as crystal-operating modes.

Table 2.1 shows the typical metal case in which modern quartz crystals are housed. There is also an illustration of a circular crystal with plated sides and electrodes. It is shown with the metal case removed. The electrodes are connected to the socket pins by means of wire leads. Dimensions for the most popular crystal cases are given in the table.

TABLE 2.1

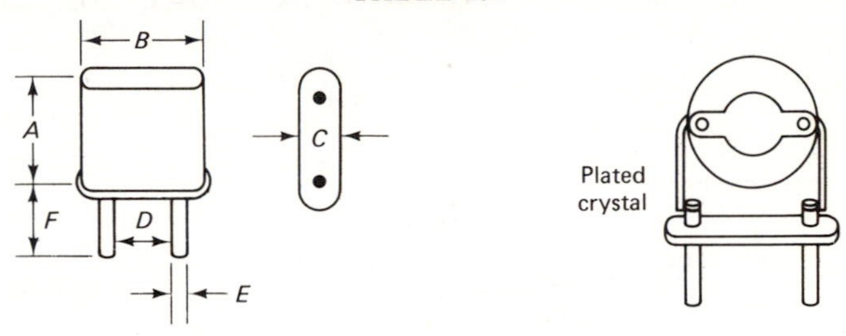

Case styles and dimensions

Case style	A	B	C	D	E	F
HC-13/U	1.516	0.721	0.316	0.486	0.050	0.238
HC-6/U	0.750	0.721	0.316	0.486	0.050	0.238
HC-17/U	0.750	0.721	0.316	0.486	0.093	0.445
HC-12/U	0.555	0.526	0.200	0.275	0.040	0.275
HC-18/U *	0.510	0.400	0.149	0.192	0.018	1.50
HC-25/U	0.510	0.400	0.149	0.192	0.040	0.250

* Wire leads, *A–F*, incl. = in., mm = in. × 25.4

2.1.1 Fundamental Crystals

A *fundamental crystal* is one that operates at the frequency for which it has been cut or ground. Presently, the upper limit for availability of this type of crystal is 21 MHz.

A crystal that oscillates on its fundamental mode should have its drive level limited to approximately 10 milliwatts (mW) at 10 MHz and lower. Between 10 and 21 MHz the drive level should be maintained below 5 mW. This will prevent excessive crystal heating, drift, and possible damage.

If operation above 21 MHz is desired, the oscillator tank can be tuned to one of the crystal harmonics and then amplified by a subsequent stage. Normally, the fifth harmonic is the upper limit for this type of operation, with the second and third harmonics being the more common ones.

2.1.2 Overtone Crystals

A better technique for operating a crystal oscillator above 21 MHz (as opposed to the harmonic operation discussed in Section 2.1.1) is that of using an overtone type of crystal. In such an event we design an oscillator that has the output tuned circuit resonant at an odd multiple of the crystal. Practically, the third and fifth overtones of the crystal are used. Figure 2.3 shows a typical overtone oscillator. $L1$ and $C1$

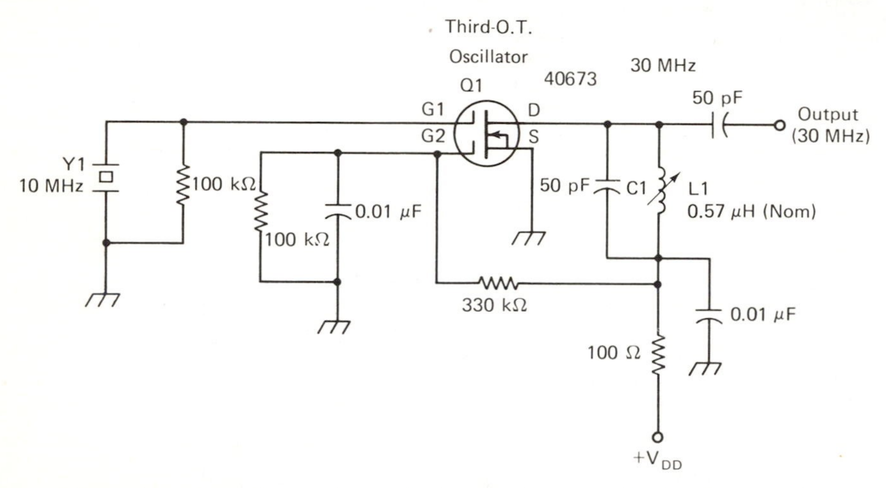

Figure 2.3 **Crystals can be made to operate on their odd overtones by supplying feedback at the desired overtone frequency.** ***C*1** **and** ***L*1** **are tuned to the desired overtone.**

form a resonant circuit at the desired output frequency. In this example we have chosen the third overtone of $Y1$. Since $Y1$ is cut for 10 MHz, the output from $Q1$ will be approximately 30 MHz, the output frequency is never an exact multiple of the crystal frequency. Also, to ensure reliable oscillator starting, the tuned circuit ($L1$ and $C1$) is resonated slightly *above* the overtone frequency.

Because the overtone frequency is always slightly above the fundamental frequency of the crystal, manufactured units are produced for the *desired* overtone frequency. The fundamental frequency of the crystal is not even considered when the crystal holder is marked for the operating frequency. Conversely, an overtone crystal can be operated on its fundamental in a different kind of oscillator circuit, but the submultiple will not be exactly related to the marked frequency. For example, if a 9-MHz fundamental crystal was operated on its third overtone, the output frequency of the oscillator would be approximately 27.010 MHz rather than precisely 27.000 MHz.

In an overtone oscillator of the type shown in Fig. 2.3, the feedback energy from the output tank must be at the desired overtone. Low drive level is as important in this circuit as in a fundamental oscillator. The lower the drive, the greater is the frequency stability. The maximum recommended drive level for any overtone crystal is 2 mW.

Overtone crystals can be manufactured in the following frequency ranges:

Overtone	MHz
Third	10 to 60
Fifth	61 to 110
Seventh	111 to 140
Ninth	141 to 160

The higher the overtone mode the more expensive the crystal becomes. Practical designers prefer to use a third- or fifth-overtone crystal and then increase the frequency by means of passive or active multiplier stages. With respect to economics, the use of frequency multipliers is usually the best approach.

2.2 PRACTICAL CRYSTAL OSCILLATORS

Although there are numerous forms of crystal oscillators we can adopt for use in our circuits, the most common ones in present use are the triode Pierce and the Colpitts. But whatever type of crystal oscillator we may elect to use, we must pay attention to the load capacitance presented by the circuit when ordering a crystal. If the crystal is not ground or etched for the specific load capacitance, the frequency of operation will not be the same as the crystal frequency. Let us take the circuit of Fig. 2.4 as an example. Here we have a typical Colpitts oscillator that

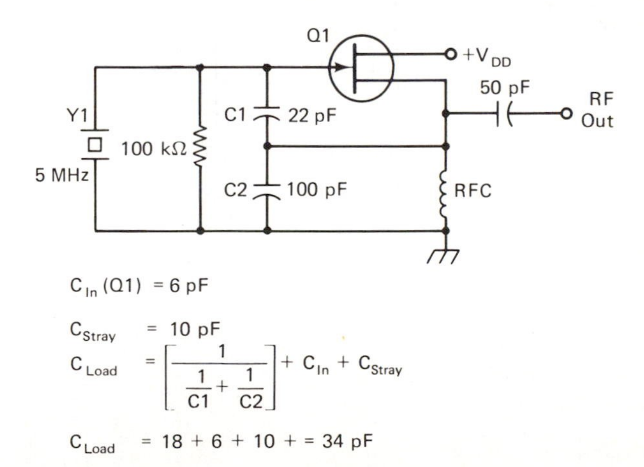

Figure 2.4 Colpitts crystal oscillator with mathematical solution of determining the crystal load capacitance.

operates at 5 MHz. The input capacitance of the FET is 6 pF. We can allow approximately 10 pF for stray circuit wiring from $Y1$ to the gate of $Q1$. Finally, we have $C1$ and $C2$ in series for the feedback circuit. By calculating all these reactances we can determine our load capacitance to be 34 pF. In ordering a close-tolerance crystal, we would tell the manufacturer to process it for a 34-pF load.

In the interest of optimum frequency stability, we may choose to use a regulated supply voltage on our oscillator. Changes in operating voltage can shift the frequency of oscillation. The higher the crystal frequency, the more pronounced is the effect. Normally, a Zener-diode regulator in the V_{cc} or V_{DD} supply line will suffice for this form of stabilization.

2.2.1 Pierce Oscillator

Figure 2.5 shows a bipolar-transistor Pierce oscillator. $Y1$ is directly in the feedback path from collector to base. The amount of feedback is

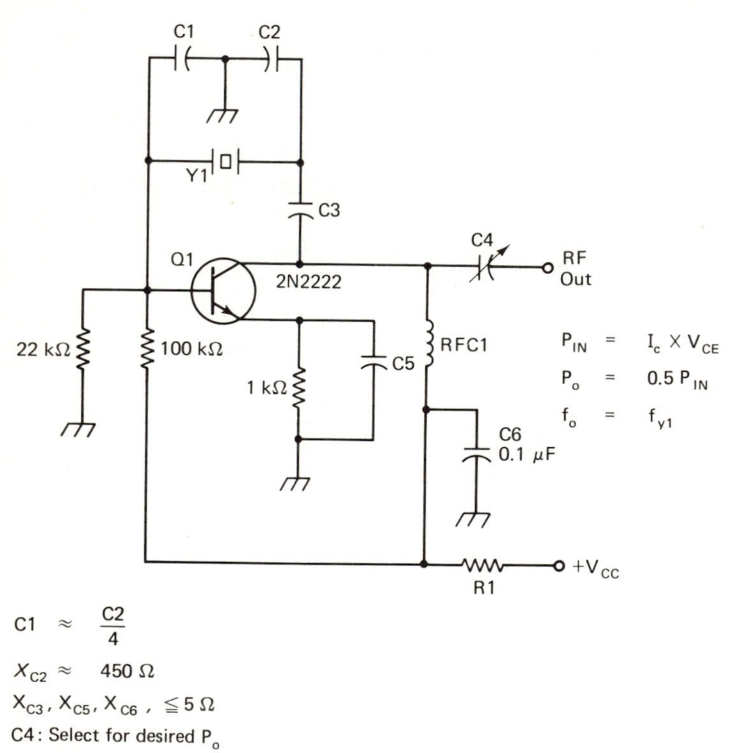

$C1 \approx \frac{C2}{4}$

$X_{C2} \approx 450\ \Omega$

$X_{C3}, X_{C5}, X_{C6}, \leqq 5\ \Omega$

C4: Select for desired P_o

X_L (RFC1) $\geqq 10\ k\Omega$

R1: To provide V_{CE} of $\frac{V_{CC}}{2}$

Figure 2.5 Pierce crystal oscillator with mathematical design solutions.

controlled by $C1$ and $C2$, which in the arrangement shown represent a capacitance in parallel with the crystal. These values must be taken into account when determining the crystal load capacitance versus the operating frequency. For most circuits the value of $C1$ will be approximately 0.25 that of $C2$. A practical value for $C2$ can be extracted from an X_c of roughly 450 Ω. Since $C3$ is merely a blocking capacitor, it should have an X_c of less than 5 Ω. The same is true of bypass capacitors $C5$ and $C6$. $Q1$ should have an f_T of five times or greater the operating frequency. Thus, a 5-MHz oscillator should contain a transistor for which the f_T is 25 MHz or higher for a common-emitter format. This is a useful rule in the interest of oscillator efficiency, even though most transistors can be made to oscillate at their f_T or slightly higher.

$R1$ of Fig. 2.5 can be chosen to yield a V_{ce} that is one half the supply voltage, V_{cc}. Although this will tend to lower the available output power of the oscillator, it will contribute to a cleaner output waveform. In all solid-state oscillators the transistor should have a V_{ceo} of at least twice the operating V_{ce}, thereby allowing the collector-emitter ac voltage to reach its full swing during oscillation.

$C4$ is shown as a variable capacitor. The amount of output coupling can be controlled by making this component adjustable. Care should be taken to prevent excessive coupling to the load, as too tight a coupling amount could prevent the oscillator from starting when V_{cc} is applied.

Figure 2.6 shows how an FET can be used as a Pierce oscillator. The advantage of this circuit is fewer components and the higher device input impedence as compared to $Q1$ of Fig. 2.5. The bipolar transistor operating class A has an input impedance of 500 to 1000 Ω in a typical case (dependent upon the biasing), whereas a JFET or MOSFET has a characteristic input impedance of megohms. The 100,000-Ω gate resistor

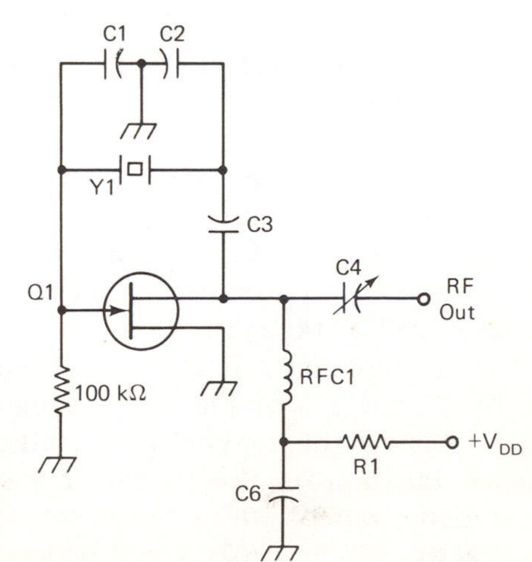

Figure 2.6 An FET can be used effectively in a Pierce crystal oscillator as shown here.

in Fig. 2.6 sets the input impedance at the value of the resistor. A dual-gate MOSFET, such as 3N211, could be substituted for $Q1$ of Fig. 2.6 by tying the gates in parallel. Alternatively, gate 2 could be biased and gate 1 used as the input gate. Approximations for a Pierce oscillator are contained in Fig. 2.5.

2.2.2 Colpitts Oscillator

Although the Pierce oscillator described in Section 2.2.1 could be viewed as a Colpitts type of oscillator, owing to the use and placement of the feedback capacitors, the more common form of Colpitts oscillator is as shown in Fig. 2.7. A classic Pierce does not show external feed-

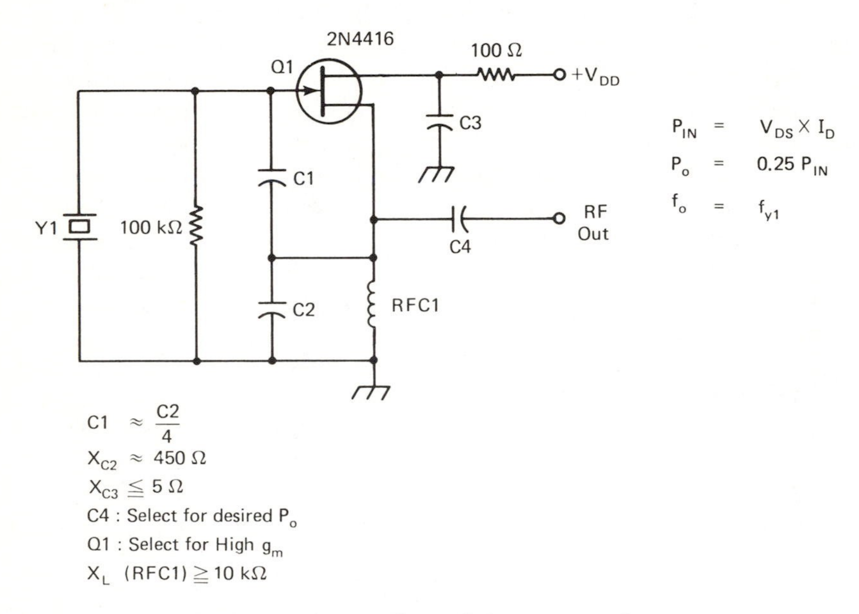

Figure 2.7 The feedback ratio for a Colpitts type of crystal oscillator is on the order of 4 : 1.

back elements, since the active device C_{in} and C_o suffice in many instances. The classic Colpitts, on the other hand, requires the use of external feedback capacitors, as shown. Some versions of this oscillator place the feedback network between drain and source and source and ground (in a FET oscillator).

In a crystal-controlled Colpitts oscillator the feedback ratio provided by $C1$ and $C2$ of Fig. 2.7 is typically on the order of 4 : 1. The drain in this circuit is bypassed for alternating current, and the signal output is taken from the transistor source terminal via $C4$. Since the characteristic impedance of the source in this circuit is low, the output voltage across $C2$ and $RFC1$ will be small. The actual voltage delivered to the load will depend on the load impedance and the value chosen for $C4$. This capacitor is ordinarily low in capacitance value (X_c is normally 450 Ω or greater). The lower the value, the less effect will be observed in oscillator frequency change with load variations, but at a sacrifice in oscillator output power. Design approximations are provided in Fig. 2.7.

2.2.3 Harmonic Oscillator

The basic form of oscillator seen in Fig. 2.7 can be applied in the circuit of Fig. 2.8 to develop harmonic output at various octaves of the

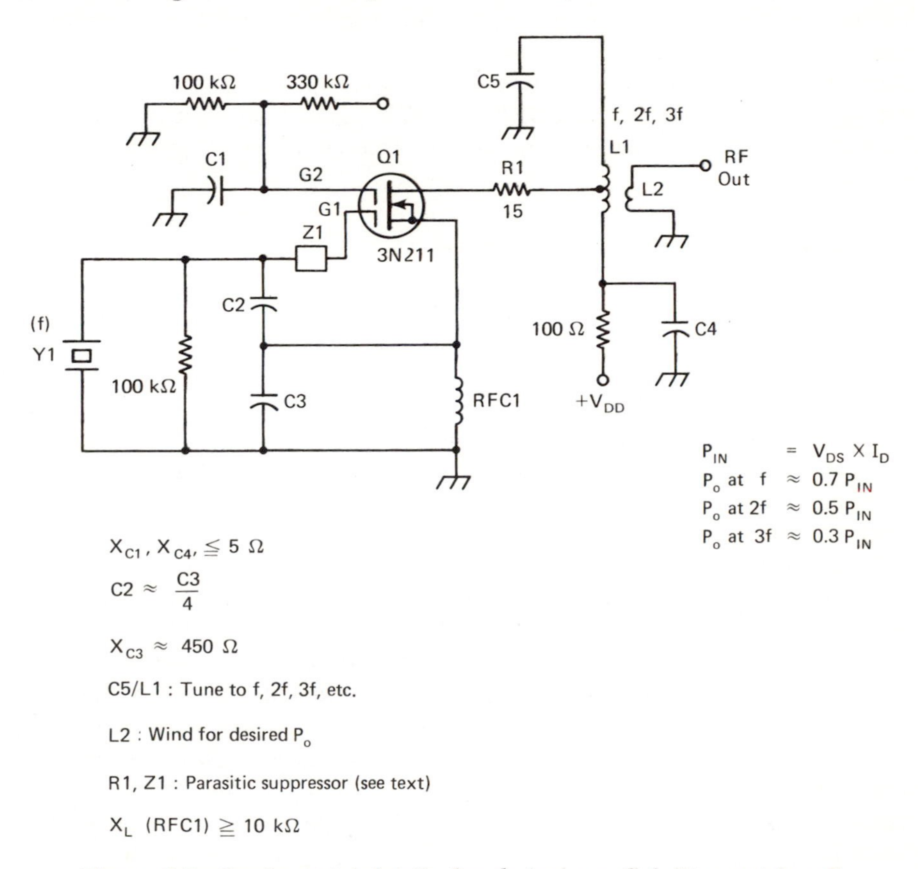

Figure 2.8 Fundamental details for designing a Colpitts crystal oscillator that uses a dual-gate MOSFET.

$Y1$ frequency. Although we have shown a dual-gate MOSFET at $Q1$, a bipolar transistor, JFET, or vacuum tube could be used as well. Alternatively, gate 2 could be connected to gate 1 to eliminate the need for the biasing resistors and $C1$ at gate 2. The effect would be similar to that obtained with a single-gate MOSFET or a JFET.

In this example we find the transistor drain tuned to the desired crystal harmonic. $R1$ presents an impedance at VHF and UHF to discourage parasitic oscillation, a frequent manifestation with this type of oscillator. A single ferrite bead at gate 1, as shown, may suffice for preventing VHF oscillations. It should have an initial permeability (μ_i) of 125 or 950 to be effective.

The drain is shown at a tap point on the tank coil, $L1$. This is done to minimize loading of the resonator by the drain, which in turn helps to preserve the tuned-circuit Q and reduces the output level of unwanted harmonics and the fundamental frequency of $Y1$. The trade-off we will experience as the tap is brought closer to the V_{DD} end of $L1$ is reduced output from $L2$, owing to an increasing mismatch between drain and output load. The loaded Q (Q_L) of the tuned circuit must be kept as high as possible to provide good rejection of spurious output energy. Devices such as the 3N211, 40673, 3N200, and equivalent high-frequency, high g_m types are well suited to this and other crystal oscillators.

2.2.4 Improved Harmonic Generation

The limitation we would experience with the circuit in Fig. 2.8 is that of having a substantial amount of fundamental and undesired harmonic currents in the oscillator output, even though the tuned circuit was of high quality. Typically, the unwanted output components will be only 10 to 15 dB lower in amplitude than the desired output signal.

A much better way to obtain harmonics of the crystal frequency is illustrated in Fig. 2.9. In this example we have tuned the oscillator drain to the crystal frequency by means of $C4$ and the primary of $T1$. The secondary of $T1$ is coupled at low impedance (50 Ω) to one winding of trifilar-wound $T2$, which is a broadband transformer wound on a ferrite core (toroid) that has a μ_i of 950. The 18-MHz signal is multiplied to 36 MHz by action of $D1$ and $D2$. If good circuit balance is maintained, the 18-MHz voltage should be suppressed by at least 30 dB at the output of the doubler (ahead of $FL1$). Suppression amounts as great as 40 dB have been obtained by the author in laboratory experiments. $D1$ and $D2$ can be matched high-frequency, small-signal diodes of the 1N914 variety. Hot-carrier diodes of the HP-2800 type are well suited to this circuit.

For nonstringent applications, it should suffice to use a single resonator at $FL1$ to further purify the 36-MHz signal. The tuned circuit would serve also as an impedance-matching network between the doubler diodes and the LO-chain (local oscillator) load. If extreme spectral purity is required at the LO output, the two-resonator filter of $FL1$ can be utilized. There will be some insertion loss, but the unwanted products will be 70 dB or greater below peak 36-MHz output.

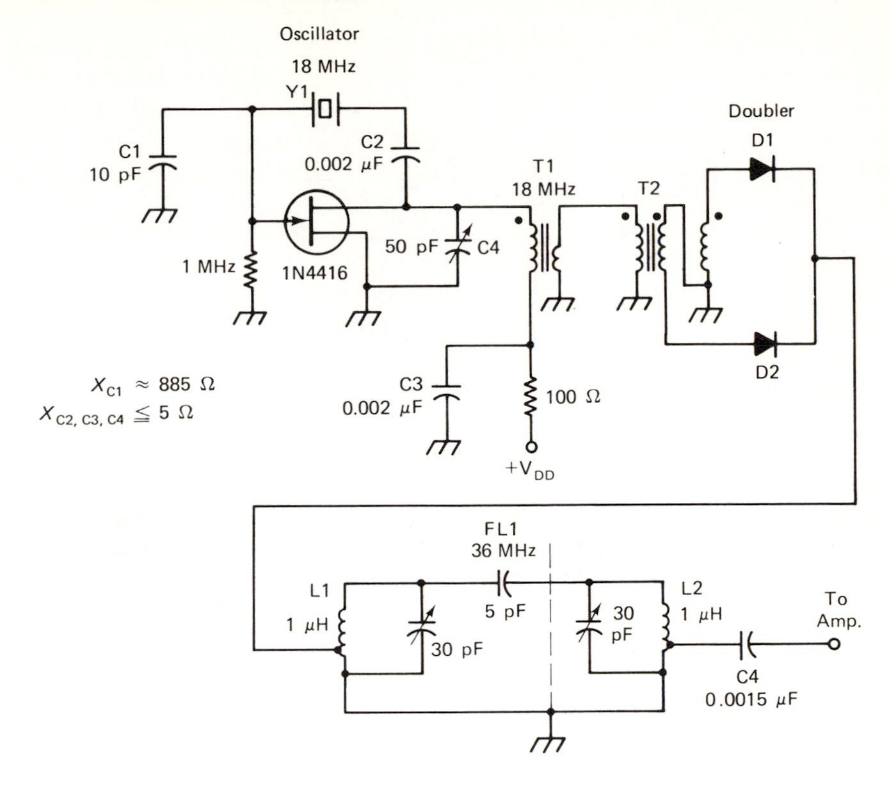

Figure 2.9 Practical circuit for a frequency-doubler circuit with spectrally clean output at 36 MHz.

The *LO* chain of Fig. 2.9 may not be suitable for applications that require a substantial amount of excitation power. The level of the 36-MHz signal can be built up easily and inexpensively by adding broadband, low-level amplifiers after *FL*1. (Data will be given later in this volume on the subject of broadband class A RF amplifiers.)

Additional frequency multiplication (even harmonics) can be accomplished by adding a broadband amplifier after *FL*1, then driving another frequency doubler of the type used in Fig. 2.9, and so on. Suitable filters would be included after each frequency doubler to ensure purity of the output waveforms.

2.2.5 Quartz Crystal Filter

We can apply what might be looked upon as the "ultimate" form of waveform purification if we use the technique shown in Fig. 2.10. *Y*1 and *Y*2 need to be on the exact same frequency in order to take advantage of the filtering properties of *Y*2, which is in series with the oscillator output. For this reason, we have included *C*1 in the design. It

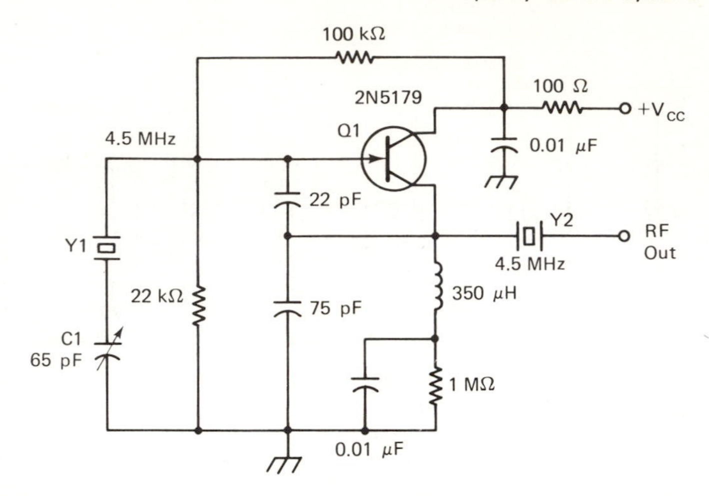

Figure 2.10 Spectral purity can be enhanced in a crystal oscillator by adding a second crystal (*Y*2) as an output filter. The center frequency of the filter must match the frequency at which *Y*1 oscillates.

can be used to "rubber" the *Y*1 frequency to that of *Y*2. This method of output filtering not only removes harmonic energy, but greatly decreases the oscillator noise output. The trade-off is in reduced *LO* power output. For most applications it is necessary to follow *Y*2 with one or two amplifier stages to develop ample injection voltage for a mixer or sufficient excitation power for a successive stage in a transmitter. The superb *Q* of the output crystal accounts for the spectral purity obtainable from this circuit.

2.2.6 Overtone Oscillator

The subject of overtone crystals and oscillators was covered earlier in the chapter, but a practical circuit is worthy of being included in this section. A JFET overtone oscillator is shown schematically in Fig. 2.11. Only the *odd* overtone of a crystal can be used for this type of oscillator. *C*1 and *L*1 form a resonant circuit at the third overtone of *Y*1 to provide feedback at the desired frequency. In practice, the drain tank will be tuned slightly higher (approximately 10%) than the overtone frequency marked on the crystal. This will ensure reliable starting of the oscillator and will minimize pulling effects respective to the operating frequency. A loaded *Q* of 10 to 15 is recommended for *C*1/*L*1. The individual *Q* values of *C*1 and *L*1 should be high to assure a sufficiently high Q_L for the circuit. A silver-mica capacitor is suitable for use at *C*1. Alternatively, *C*1 could be a 60-pF trimmer and *L*1 an air-wound inductor. If the load that follows *C*2 is of low impedance, *C*2 should be small in capacitance value to prevent loading and *Q* degradation of the drain

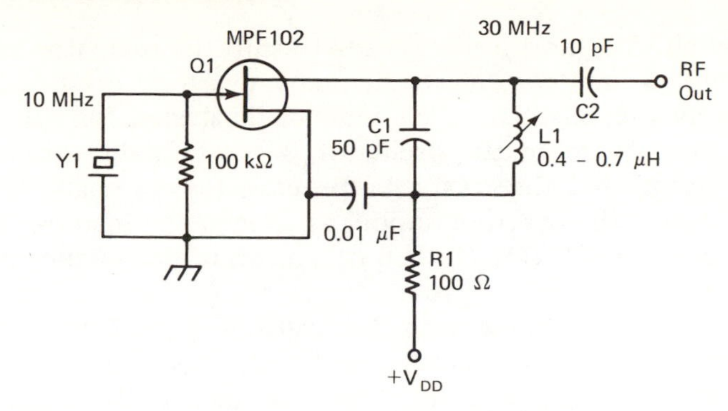

Figure 2.11 Practical circuit for a JFET 30-MHz overtone oscillator.

tank. Link coupling can be used in place of capacitive coupling. The output link would be situated at the $R1$ end of $L1$ if this were done.

2.2.7 Bimode Oscillator

Occasionally we have need for a single oscillator that can be used with fundamental and overtone crystals. A simplified version of a two-frequency, bimode JFET oscillator is given in Fig. 2.12. $S1$, a low-capa-

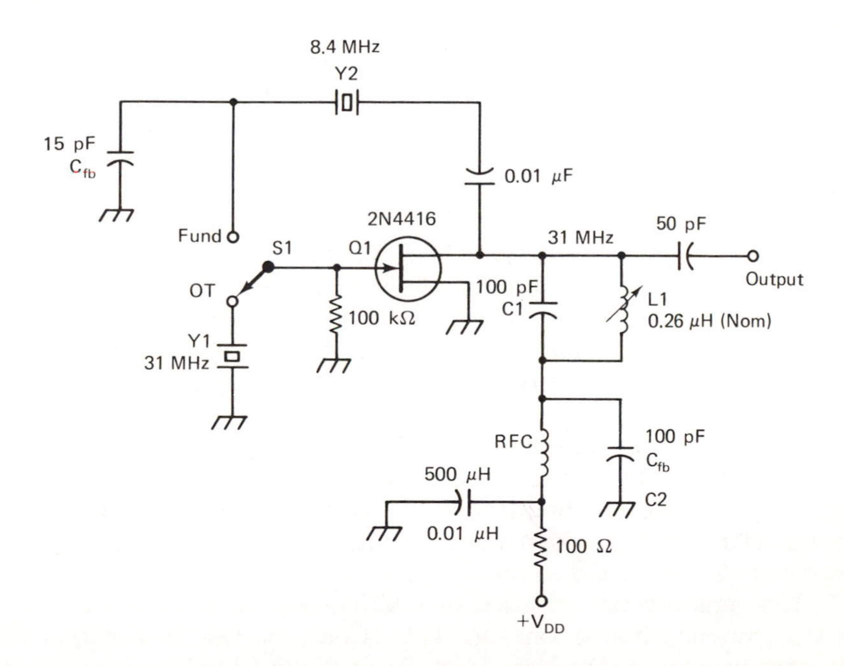

Figure 2.12 Method for obtaining fundamental and overtone operation from an oscillator that accommodates two operating frequencies.

citance switch with short leads, is used to shift the oscillator mode from *fundamental* to *overtone*, although many overtone crystals will function properly if connected in the manner illustrated for $Y2$. However, the load capacitance of the oscillator is substantially greater in that mode, requiring that the crystal be processed accordingly. With $S1$ of Fig. 2.12 set for the overtone mode, as shown, the load capacitance is approximately 16 pF ($Q1$ C_{in} = 6 pF and stray capacitance is approximately 10 pF).

$L1$ and $C1$ are resonant at the overtone frequency of $Y1$. When $S1$ is changed to the fundamental oscillator mode, $C2$ becomes part of the oscillator feedback network. The reactance of the drain tank ($L1$ and $C1$) is not significant enough at 8.4 MHz to impair circuit performance. Several crystal frequencies (OT (overtone) and fundamental) can be accommodated by this circuit if additional switching is provided at $S1$, and if the various overtone tuned circuits are switched at the $Q1$ drain. Diode switching, which will be discussed later in this book, could be employed in place of mechanical switching.

2.3 VARIABLE-FREQUENCY CRYSTAL OSCILLATOR

A variable-frequency crystal oscillator (VXO) is useful when it is necessary to shift the crystal-oscillator frequency several kilohertz away from its crystal-controlled frequency. In essence, we can "rubber" the crystal below the frequency for which it is ground or etched. This is done by adding series X_L and parallel X_C, as shown in Fig. 2.13.

If stray capacitance is kept minimized in this circuit, we should be able to obtain a frequency shift of 10 to 15 kHz at 7500 kHz with the circuit constants shown. The actual amount of frequency change will depend on the characteristics of the crystal: some crystals will rubber better than others, depending on the cut of the crystal and the capacitance introduced by the crystal holder and its mounting hardware. AT-cut crystals are considered the best for VXO use. Overtone crystals that are operated on their fundamental frequencies yield good results in VXOs also.

The maximum amount of frequency shift is a function of frequency. The available change is approximately halved each time the frequency is made one octave lower; that is, if a 10-MHz crystal could be shifted 15 kHz in a given VXO circuit, a 5-MHz crystal could be rubbered only 7.5 kHz, and so on.

If several crystals are used in a VXO, the selector switch must be of the low-capacitance variety. This is because the greater the stray (shunt) capacitance, the less effect $C1$ of Fig. 2.13 will have on the frequency. In an ideal VXO there would be no capacitance other than the tuning capacitor, and it would exhibit zero capacitance with its plates

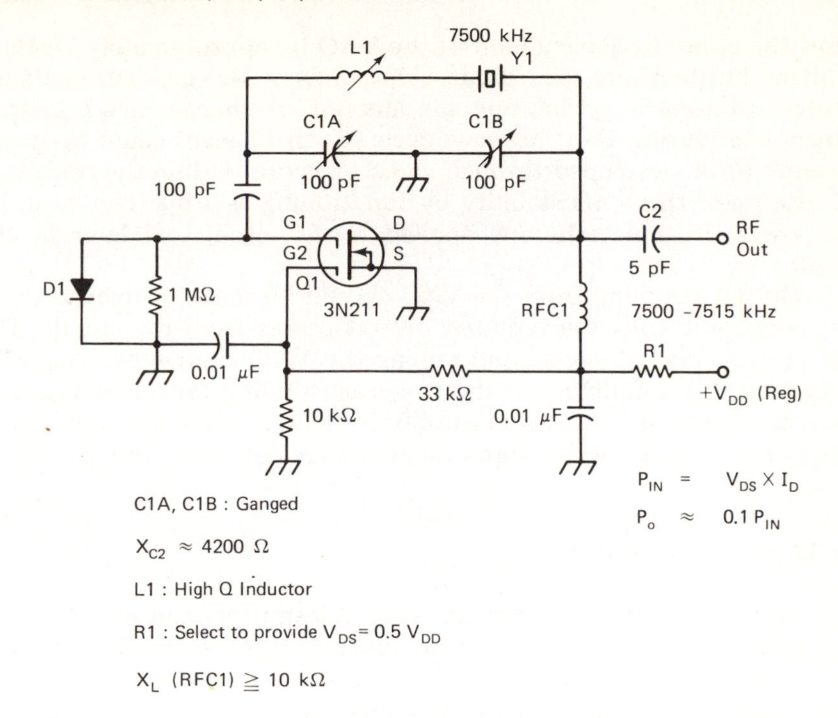

Figure 2.13 Variable-frequency crystal oscillator (VXO) that offers limited frequency change (15 kHz at 7.5 MHz) and high stability.

unmeshed. Similarly, the active device ($Q1$) would have no inherent C_{in}.

The old-style FT-141 crystal holders are generally unsuitable for VXO applications because they contain a substantial amount of capacitance between the metal pressure plates. The HC-6/U holder is much better for VXO use because the plated crystals have small contact areas, comparatively, and they are supported on small wire leads. The crystal case should not be grounded in a VXO.

A VXO that is followed by a frequency-multiplier chain can provide 100 kHz or more of frequency change at VHF with good stability. For example, an 8.333-MHz VXO output that exhibits 12 kHz of frequency shift can be multiplied 18 times to 150 MHz where the effective frequency change would be 216 kHz. This technique is useful for tunable VHF receivers that require high orders of stability. Furthermore, the output spectrum of a properly designed VXO is quite clean in terms of noise and spurious products when followed by low-level amplifiers, such as class C 2N5179s that are filtered for harmonic supression.

$D1$ of Fig. 2.13 limits the positive sine-wave excursion of $Q1$ to limit the transconductance of the dual-gate MOSFET. The advantage is that the internal capacitance of $Q1$ is minimized during operation. This raises the upper frequency limit of the VXO by approximately 1 kHz at 8 MHz. Furthermore, the diode helps prevent the generation of unwanted harmonics by limiting the amount of change in $Q1$ internal capacitance during the sine-wave cycle. Such changes cause harmonic currents to be developed through varactor action within the transistor. $D1$ also helps the VXO stability by functioning as a bias regulator. We will see this same technique applied to LC oscillators later in this chapter.

Output coupling from the VXO should be made as light as possible, consistent with the required output power from the circuit. This will prevent load changes from pulling the VXO frequency. The VXO voltage can be amplified to the necessary driving power or injection levels by means of small-signal amplifiers. Such a solution is not complicated or expensive. Design approximations are contained in Fig. 2.13.

2.4 LC RF OSCILLATORS

The tunable master oscillators in most transmitters and receivers that are not synthesized contain stable LC types of VFOs (variable-frequency oscillators). It is not difficult today for us to approach closely the short- and long-term frequency stability of a crystal oscillator while using L and C components for frequency control. Perhaps the most common of the VFOs employs the Colpitts form of oscillator, although some designers still find the Hartley oscillator entirely suitable for their equipment-performance needs.

The fundamental objectives are *frequency stability* (electrical and mechanical), *purity of output waveform*, and *linearity of tuning*. The latter is important mainly in systems that use analog display of the operating frequency: it is much more convenient to have dial calibration that is linear than to have part of the tuning range compressed at one end of the dial scale while the other end is spread out. Various physical configurations have been developed for variable capacitor (midline, straight-line, etc.) plates to achieve linearity of readout, but the best technique is to have a lot of C in the VFO tank, and then use a small amount of variable capacitance to change the frequency. In that manner, the percentage of capacitance change is quite small with respect to the total effective capacitance. The linearity of tuning then becomes acceptable for most designs.

2.4.1 Component Considerations

Our choice of components for the frequency-determining part of the VFO circuit is necessarily stringent if we are to design a circuit that is stable. These recommendations will be useful to all persons who become involved with *LC* oscillator fabrication.

Variable capacitors. The main tuning capacitor should be of the double-bearing type (rotor bearing at each end). This will help ensure positive electrical contact between the rotor and the circuit point to which it connects. Also, the double-bearing capacitor offers mechanical rigidity that discourages the rotor from shifting on its axis during mechanical stress. Such shifting changes the relationship of the stator and rotor plates, thereby changing the capacitance at a given dial setting. Brass plates with cadmium or silver plating are more temperature stable than are aluminum plates. This suggests strongly that the latter should be avoided in VFOs.

Inductors. Slug-tuned VFO coils should be well built in order to resist changes from mechanical stress and the expansion and contraction brought about by large changes in ambient temperature. The coils should be built so that the adjustable slug can be locked firmly in position once the oscillator has been calibrated.

Coils with litz wire windings are preferred in the interest of increased coil *Q*, although enamel-covered magnet wire is satisfactory for many applications.

Toroid or pot cores are not recommended for use in an *LC* oscillator tank unless the anticipated temperature environment of the composite equipment is to be relatively constant. Large magnetic cores are subject to a considerable amount of permeability shift with changes in temperature, which in turn can cause severe problems with long-term drift. Ideally, we would use no core material other than air. This limitation would place constrictions on the equipment size, however, as air-wound inductors are necessarily quite large when built for rigidity. The practical approach to oscillator-coil selection is the employment of a high-quality slug-tuned inductor that has its slug just into the end of the coil when the tuned circuit is adjusted for its normal operating range. This will minimize the effects of permeability change, since the amount of iron or ferrite in the field of the coil will be small compared to what it would be if a toroid core or a pot core were used.

The completed oscillator coil should be doped heavily with a low-loss compound such as Polystyrene Q Dope to prevent the coil turns from moving. This will also prevent moisture from causing the apparent inductance to increase as the humidity rises.

Oscillator coils should be mounted at least one coil diameter away from nearby conducting objects, such as the shield enclosure. This procedure will prevent degradation of the tuned-circuit *Q* and will greatly reduce the effects of vibration with respect to momentary frequency instability.

LC oscillator coils should be shielded to prevent stray coupling to adjacent circuits, which could cause frequency instability through the introduction of unwanted RF energy. A properly shielded coil will be relatively immune to frequency shifts of rapid nature when momentary changes in ambient temperature occur, such as air drafts.

Fixed-value capacitors. Most magnetic core materials cause a positive drift characteristic when used in an *LC* oscillator. For this reason it is necessary to utilize capacitors that have a negative drift trait. Unfortunately, the amount of drift compensation is usually a cut-and-try proposition, because all the performance characteristics of the oscillator components are not known. A practical oscillator often has a combination of capacitor types, zero coefficient and negative coefficient. Up to 10 MHz it is practical to use polystyrene capacitors. They exhibit a slight negative-temperature coefficient, making them ideal for use with magnetic cores that move in the positive direction. Silver-mica capacitors can be used, but their individual drift characteristics can vary from negative to zero to positive, even though a group of them is selected from one manufacturing run. If they are used, they should be temperature graded over a 1-hour or greater period in a temperature chamber while monitoring the capacitance value with a digital capacitance meter. The test-chamber temperature can be varied from, say, 50° to 100°F. Dipped silver micas, once graded, are excellent for use in oscillators because they exhibit high *Q* and are resistant to moisture and chemicals. The same is true of polystyrene capacitors.

Printed-circuit boards. At medium frequency and higher, the *LC*-oscillator circuit board plays a significant role in circuit stability. This is because the board dielectric and the copper conductors form unwanted capacitances that can change markedly in value with temperature and circuit-board physical stress. This is especially true when double-sided boards (copper on both sides) are employed. For this reason it is prudent to use only single-sided PC board material in the frequency-determining part of the oscillator circuit, thereby avoiding the probable effects of parasitic capacitances. Furthermore, high-quality glass-epoxy PC boards of reasonable mil thickness are recommended. This will help ensure mechanical stability while preventing *Q* degradation in key parts of the circuit.

Connecting leads. Wire leads that join RF points on an oscillator circuit board to variable capacitors, switches, and jacks are a frequent cause of thermal and mechanical instability. This is because they may become absorbed into one of the critical inductances of an oscillator circuit, thereby becoming a part of the operating inductance. If they move significantly under mechanical stress, the oscillator frequency will change. If they expand and shift position as the ambient temperature rises, they can aggravate the long-term drift problem. Additionally, the interconnecting leads can degrade the Q of an oscillator coil to which they are attached if they represent a significant part of the coil inductance (one-tenth the coil inductance or greater). Short rigid leads, routed away from nearby conducting objects, will aid in preventing the foregoing difficulties.

2.4.2 Hartley Oscillator

An *LC* oscillator functions in essentially the same manner as the crystal oscillators we discussed earlier in the chapter. That is, the oscillator is essentially a tuned amplifier in which a portion of the output voltage is routed to the input port at a phase difference of 180°, causing oscillation to occur. This positive feedback can be obtained through mutual inductance, as in Fig. 2.14(a), or through capacitive coupling from the output port to the input port when separate tuned circuits are used at the oscillator input and output.

In the circuit of Fig. 2.14(a) we can control the feedback amount by means of $C1$ or $C3$, although $C3$ can be considered mainly as a blocking capacitor to isolate V_{DD} from the gate of $Q1$. Some capacitive feedback exists because of tuning capacitor $C2$, and the level of feedback voltage will vary with the setting of $C2$. We must consider this condition if we desire uniform voltage or power output from the oscillator. Additional feedback capacitance exists within $Q1$ from drain to gate.

A variation of the Hartley circuit is shown in Fig. 2.14(b). In this example we find $C2$ and $C3$ as fixed-value components. Tuning is effected by virtue of $C1$. If $C2$ and $C3$ are large in capacitance and $C1$ is relatively small, the setting of $C1$ will have little effect on the feedback ratio. A ratio of 10 : 1 or greater is recommended for $C1$ to $C2$, with the maximum value of $C1$ (plates meshed) representing a part of $C2$. Hence, $C3$ is specified slightly larger than $C2$. Feedback can be controlled by the ratio of capacitance of $C2$ and $C3$, but this circuit operates well with identical values for $C2$, and $C3$ if the Q of $L1$ is high. Standard parts values are listed in Fig. 2.14(b) to make duplication easy.

Perhaps the most popular form of the Hartley oscillator is that of Fig. 2.14(c). Here we have a tapped-coil configuration that permits easy control of the feedback voltage. Feedback is taken from the source of $Q1$ and routed to the gate via $L2$. The desired 180° phase relationship is possible by virtue of the drain current flowing through the source. This allows a portion of the output current to flow through the tuned circuit in phase with the gate current. In this example we can consider the

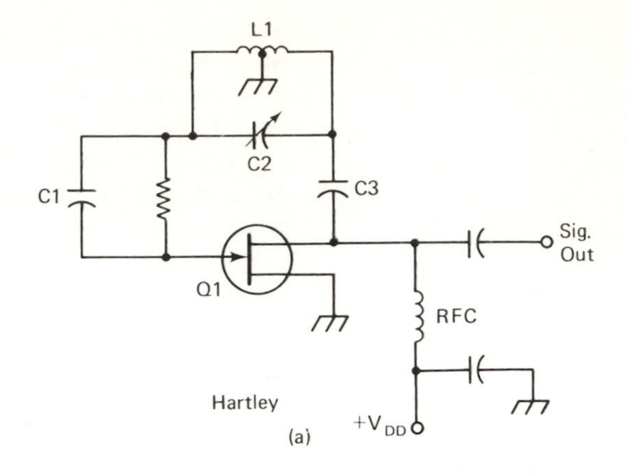

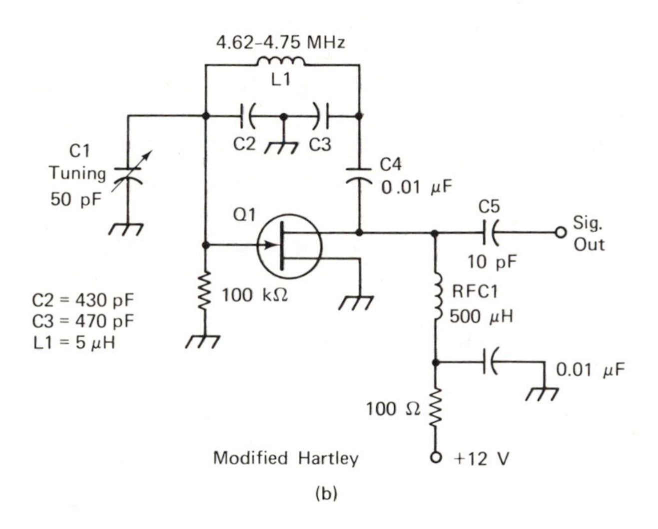

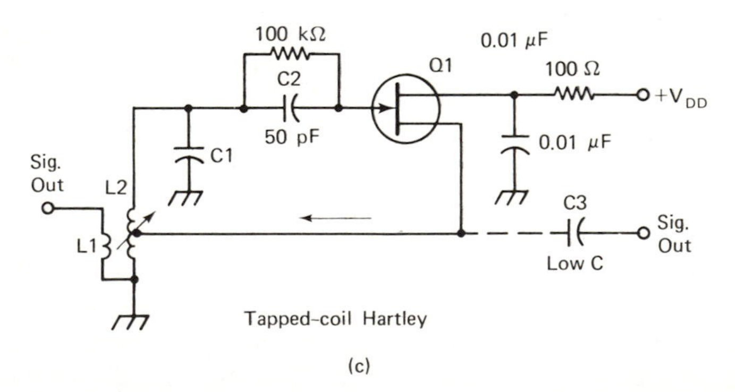

Figure 2.14 (a) Hartley oscillator, and (b) modified Hartley oscillator; (c) tapped-coil Hartley oscillator. Its feedback is determined by the placement of the coil tap on $L2$.

drain and gate as being connected to opposite ends of the tuned circuit [as in Fig. 2.14(a) and (b)], at which points the voltages are of opposite phase. If the source tap is placed at either end of $L2$, the feedback will be virtually zero. At the center of $L2$ the feedback will be maximum. For circuits that contain a high-Q tank circuit, and for which the $Q1$ g_m is reasonably high, the source tap can be placed approximately one-fourth the way up from ground on $L2$. It should not be necessary to elevate the tap point beyond one-third the total coil turns above ground. The smaller the value of $C2$, the lighter will be the loading on the resonator, and hence the higher the Q. This will enhance the purity of the output voltage in terms of noise and harmonic currents.

Output from the circuit of Fig. 2.14(c) can be taken by means of $C3$, using the smallest value of capacitance possible consistent with adequate output voltage. This will minimize frequency changes resulting from changes in load conditions. An alternative method for extracting output from the oscillator is by adding a secondary winding to $L2$. It should be placed over or below the grounded end of $L2$, and the number of turns can be adjusted to provide the desired output voltage.

2.4.3 Colpitts Oscillator

The fundamental form for the Colpitts oscillator is seen in Fig. 2.15(a). We can see that it is similar to the Hartley oscillator just discussed. Feedback is provided by $C1$ and $C2$, which generally are of identical capacitance value. Since the drain current of $Q1$ flows through the transistor source, the required 180° phase shift necessary for oscillation occurs. The source obtains its dc ground through $RFC1$, whereas in the Hartley circuit of Fig. 2.14(c) the source return is through $L2$.

A variation of the Colpitts oscillator that is popular with designers who use bipolar transistors is shown in Fig. 2.15(b). In this example we find the transistor base bypassed for alternating current, and the collector is above ac ground. $C1$ and $C2$ comprise the feedback elements in addition to the intrinsic capacitance of $Q1$. Output coupling can be as shown with $L2$, or a small-value capacitor can be connected to the top of $L1$ for sampling the oscillator output energy. The tank circuit can be tuned by placing a variable capacitor across $L1$. A smaller tuning range (bandspread) will result if the variable capacitor is connected in parallel with $C2$. $Q1$ in this circuit should have a high f_T respective to the operating frequency. Ideally, the f_T will be at least ten times the operating frequency. This will ensure easy starting and reasonable feedback requirements.

2.4.4 Series-Tuned Colpitts

A variation of the circuit given in Fig. 2.15 is the series-tuned Colpitts oscillator. This type of oscillator is known as the "series-tuned

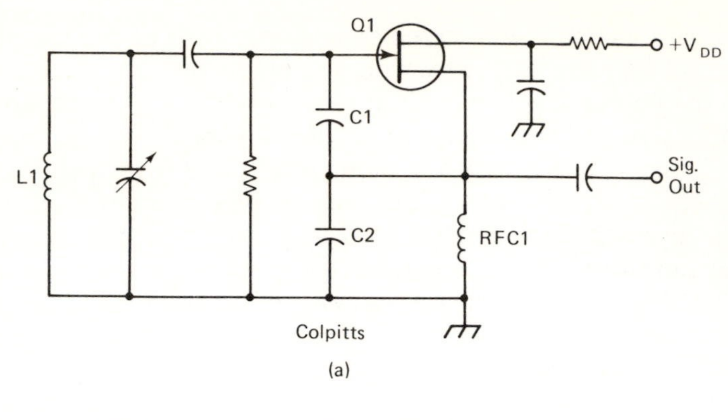

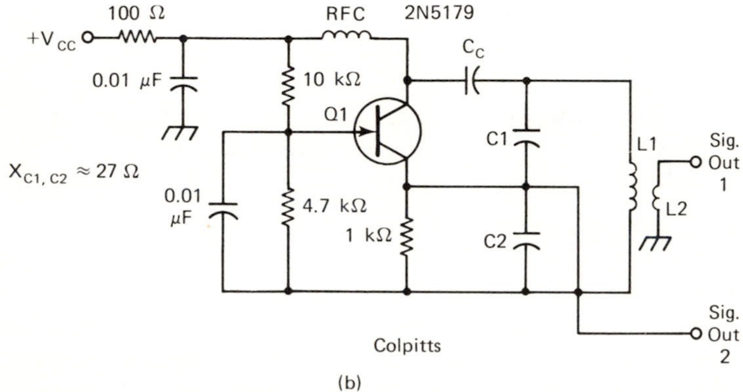

Figure 2.15 (a) Basic Colpitts *LC* oscillator, and (b) a practical version of the same type of oscillator.

Clapp" oscillator by some designers. The circuit in its basic form is shown in Fig. 2.16. It differs from the parallel-tuned Colpitts only by virtue of the arrangement of $L1$ and $C1$.

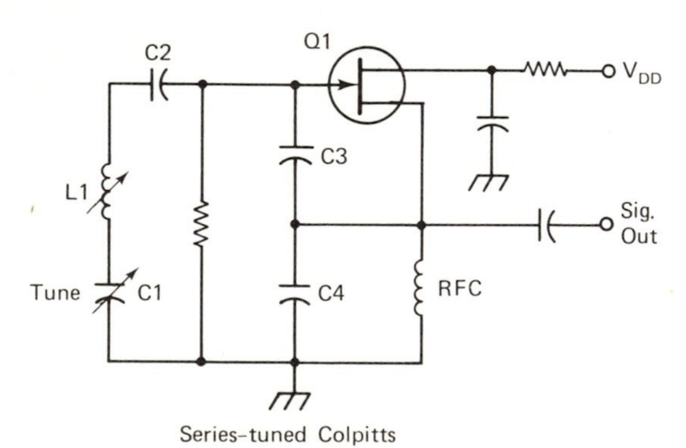

Figure 2.16 High stability is possible when using this series-tuned Colpitts *LC* oscillator.

There are two distinct advantages associated with the series-tuned version: a large frequency change is possible as compared to the parallel-tuned arrangement, assuming $C1$ in both cases has an equivalent capacitance range. This is significant when a physically small variable capacitor is desired in compact equipment. The second advantage is realized at frequencies above approximately 5 MHz, where with a parallel-tuned Colpitts oscillator the inductance of $L1$ may necessarily become an impractical value. This factor becomes more pronounced as the operating frequency of the VFO is increased toward 30 MHz. This effect can be minimized to some extent by making $C2$ as small in value as practical, consistent with reliable oscillation of $Q1$. The small amount of inductance at $L1$ (in a parallel tank) results from the high amount of shunt capacitance in the feedback capacitors, plus that of the variable capacitor, $C1$.

By adopting the circuit of Fig. 2.16, it is possible to use substantially more inductance at $L1$. Specifically, the circuit of Fig. 2.15 has a high C-to-L ratio, whereas the oscillator of Fig. 2.16 has a high L-to-C ratio. In either case it is important to ensure a high value of Q for the oscillator tank circuit. This will result in minimum oscillator noise bandwidth and improved output waveform purity.

Still another benefit of series tuning the tank is related to the increased inductance of $L1$: when the required inductance of $L1$ is very low, say, less than 2 microhenries (μH), the circuit leads between $L1$ and $C1$ and between $L1$ and $C2$ can comprise a significant part of the total effective inductance. In a situation of that kind we may experience Q degradation and problems with frequency stability. Furthermore, stress on the VFO circuit board and shield compartment can have a marked effect on the effective inductance of the tuned circuit, thereby causing mechanical instability as well as long-term thermal instability. The connecting leads become a less significant part of the inductance when $L1$ is of a relatively high value. The problems we have just treated are compounded when the VFO covers several HF ranges and employs switched coils at $L1$. Diode switching is preferable to mechanical switching if we are to minimize the inductive effects of the coil leads.

A set of component values have been assigned to the series-tuned oscillator of Fig. 2.17 to demonstrate typical values for a frequency range of 8100 to 10,335 kHz. Figure 2.17(b) shows us how the total effective capacitance of the tuned circuit is determined. C_{stray} represents 5 pF of $Q1$ input capacitance, plus 5 pF of stray circuit capacitance. We have ignored the effective minimum C of the 50-pF tuning capacitor for the purpose of simplifying the discussion. In a typical case it would be somewhere between 3 and 10 pF, depending on the physical format of the component and the mounting technique used.

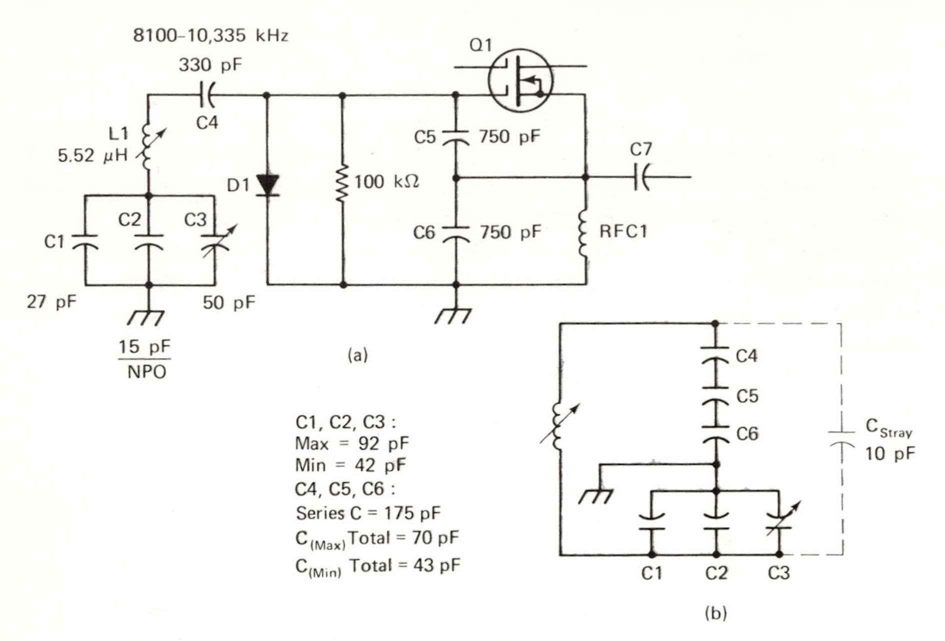

Figure 2.17 Method of determining the total effective capacitance in a Colpitts *LC* oscillator.

The effective change in capacitance at the lower end of $L1$ is 50 pF, by virtue of $C3$. Because we have included fixed-value capacitors $C1$ and $C2$, they set the minimum capacitance in that part of the circuit at 42 pF. When the plates of $C3$ are fully meshed, we have a maximum capacitance of 92 pF in that part of the circuit. $C1$ through $C6$ can be considered in series, as shown in Fig. 2.17(b). This combination of capacitances, plus C_{stray}, yields a total maximum tank capacitance of 70 pF, with the minimum value being 43 pF. The 5.52-μH inductor at $L1$ will provide the tuning range specified.

We have neglected mentioning $C7$, the output coupling capacitor. It will have a small effect on the tuning range if its value is low. It can be considered in series with the load capacitance of the circuit that follows $Q1$. The net value of $C7$ and the load capacitance can then be added to that of $C6$ as a parallel capacitance before calculating the effective capacitance range of the operating oscillator. These minor effects can be compensated for in an empirical fashion if we make $C1$ a trimmer capacitor. Then, with $L1$ and $C1$ as variable elements in the circuit, we can juggle their settings to provide exact frequency coverage over the chosen range.

To illustrate the difference in $L1$ value between the circuit of Fig. 2.15 and the one in Fig. 2.17, the coil inductance would be approximately 1.6 μH for the parallel-tuned circuit, as opposed to 5.52 μH in the series-tuned example. This assumes that all values of capacitance remained unchanged. Furthermore, we would lose roughly 1 MHz of tuning range, 8100 to about 9100 kHz, by adoption of the parallel-tuned Colpitts oscillator.

In the interest of frequency stability we have used three capacitors at the bottom end of *L*1. It would be beneficial to use even more. The rationale is that the circulating RF current is distributed through several capacitors, rather than one or two. This reduces the internal heating (however miniscule) of each capacitor and enhances the long-term frequency stability. A similar approach can be used at *C*4, *C*5, and *C*6 if high orders of stability are required.

*D*1 of Fig. 2.17 aids stability by functioning as a bias clamp on the gate of *Q*1. It prevents high positive voltage peaks during the sine-wave cycle, which in turn restricts the transconductance of *Q*1. This prevents large changes in *Q*1 junction capacitance, thereby minimizing harmonic generation and internal heating of *Q*1. If a JFET is used instead of the MOSFET shown, the gate-source junction will serve as *D*1, provided the source is grounded. If the source is above ground, however, the clamp diode is recommended.

*D*1 of Fig. 2.17 can also improve the signal-to-noise ratio of the oscillator by preventing gate-to-channel conduction in the JFET. Generally, this concept can be applied to the other oscillator circuits shown in this chapter also. A Schottky diode should be used for this purpose if we are to obtain good results.

2.4.5 A Practical Colpitts Oscillator

Figure 2.18 contains the circuit of one of the author's laboratory models for testing the Colpitts VFO. The reactance values listed in the figure can be used to extrapolate other operating frequencies. Calculations for the tuning range and effective minimum and maximum *C* can be based on the data given for the circuit of Fig. 2.17.

A 3N211 dual-gate MOSFET was chosen for the circuit because of its extremely high g_m. A variety of similar devices could be used (40673 for one), but the high g_m was desired in the interest of high performance with a minimal amount of feedback requirements. The device characteristic we are discussing is frequently termed g_{fs}, which equates to the device g_m.

A 10-pF NPO capacitor is used at *C*1 to compensate for the positive-drift characteristic of the core material in *L*1. Up to at least 10 MHz we can use polystyrene capacitors at *C*2, *C*4, *C*5, *C*6, and *C*7 to enhance stability. They have a small negative coefficient, which is beneficial in countering the positive drift of *L*1. Silver-mica capacitors will yield good *Q* and reasonable temperature stability in place of the polystyrene units, but they may have positive or negative drift, depending on the production run from which they came. The cut-and-try method

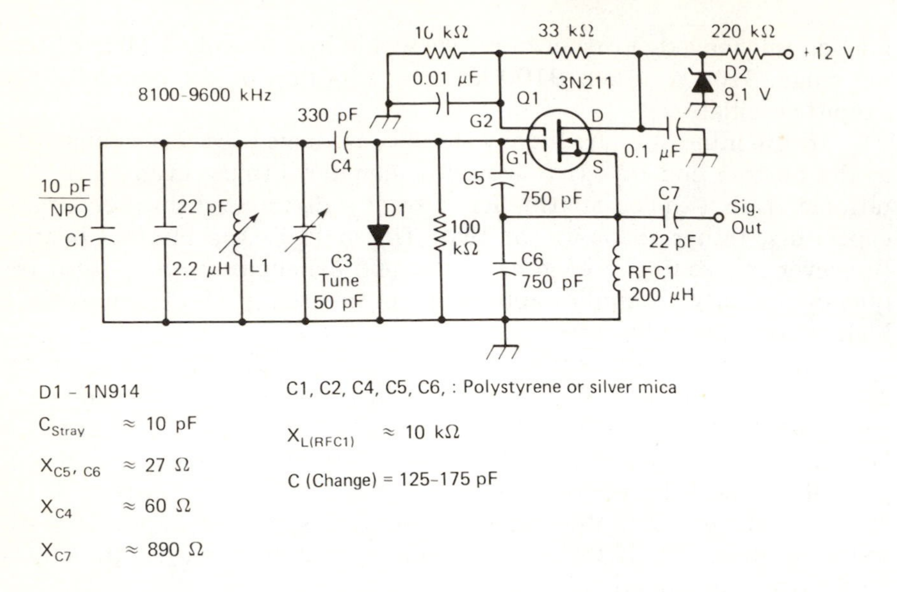

Figure 2.18 Practical example of a stable, parallel-tuned Colpitts VFO with approximations for the component values.

of obtaining good stability is almost mandatory when using silver micas, unless they have been temperature graded before placing them in the circuit.

$D1$ in this circuit is a small-signal, high-speed diode of the 1N914 family. A hot-carrier diode would work well in that part of the circuit also.

Frequency stability of the VFO in Fig. 2.18 was entirely acceptable for most receiver or transmitter applications. From a cold start to a period 3 hours later the drift was 93 Hz upward. Short-term stabilization occurred approximately 5 min after turn-on, accounting for 50 Hz of the total drift. Long-term stabilization took place in 1 hour. Thereafter, the operating frequency ramped slowly upward and downward over a 10-Hz range. Testing was done in a sealed compartment at 25°C. A high-quality J. W. Miller Co. 43-series slug-tuned inductor was used at $L1$, and $C1$ was a double-bearing unit with plated-brass vanes. Output across a 1000-Ω resistive termination was 2 V peak-to-peak, as measured with a Tektronix model 453 scope.

Gate 2 of $Q1$ can be tied directly to gate 1 if desired. This would eliminate the need to bias gate 2. The transistor would then operate as a single-gate MOSFET.

A series-tuned version of the VFO in Fig. 2.18 was built and evaluated at 2 MHz. The tuning range was 2.0 to 2.2 MHz. Frequency stability was so good that no drift could be measured over a 3-hour period. It is likely that improved stability would result with the 8.1-MHz version of Fig. 2.18 if series tuning were employed.

2.5 VVC TUNING

Voltage-variable-capacitance (VVC) diodes are suitable for use in changing the operating frequency of a VFO. These diodes are also called Varicaps, Epicaps, and by other trade names. In essence, VVC diodes are high-*Q* junction semiconductors that are used in place of mechanical variable capacitors. The junction capacitance is varied by applying different amounts of reverse bias to the diode. A positive potential is placed on the diode cathode, and as the dc level is changed, so is the effective capacitance. An illustration of the capacitance variation with respect to various dc voltage levels is given in Table 2.2. The listed values are for a Motorola plastic-cased MV109 tuning diode. We can see from the table that as the reverse bias is increased the junction capacitance decreases. As this is happening, the diode *Q* increases.

There are many types of VVC diodes. Some have several hundred picofarads of capacitance, while others have a maximum capacitance of only a few picofarads. Also, they are rated for a variety of operating frequencies.

The primary shortcoming of VVC diodes as tuning elements is temperature stability. Each time we introduce another semiconductor junction in the frequency-determining part of an oscillator circuit, we

TABLE 2.2 Capacitance Variation with Respect to DC Voltage Levels

Motorola MV109 (Approx. Values)	
V (Reverse)	C(pf)
30	5
20	7
15	8
10	12
9	13
8	15
7	16
6	18
5	21
4	24
3	30
2	35
1	39

Q at 50 MHz (O V) = 300
Q at 50 MHz (12 V) = 3000

add another possible cause of frequency instability. This is because as the ambient temperature around the diode changes so does the junction capacitance, however slight that change may be. RF currents passing through the diode will also cause some heating of the junction.

Other types of devices can be used as VVC tuning diodes. For example, we might employ a 1N914 diode if only a small capacitance change were required. Two or more 1N914s could be used in parallel to increase the capacitance. Alternatively, a pair of *npn* bipolar transistors will function as a VVC tuning system when connected as shown in Fig. 2.19. The base-collector junction serves as the tuning diodes at *Q*1 and *Q*2. The emitters are not used. Transistors such as the 2N2222A and 2N3053 are among those that lend themselves to this application. The amount of junction-capacitance change will vary with the type of transistor used. *C*1 is in parallel with *L*1 to establish the center of the operating frequency range, while *R*1 is set at midrange. *L*1 must be grounded in order to provide a dc return for *Q*1 and *Q*2. By employing the back-to-back diode format shown in Fig. 2.19, we ensure good waveform

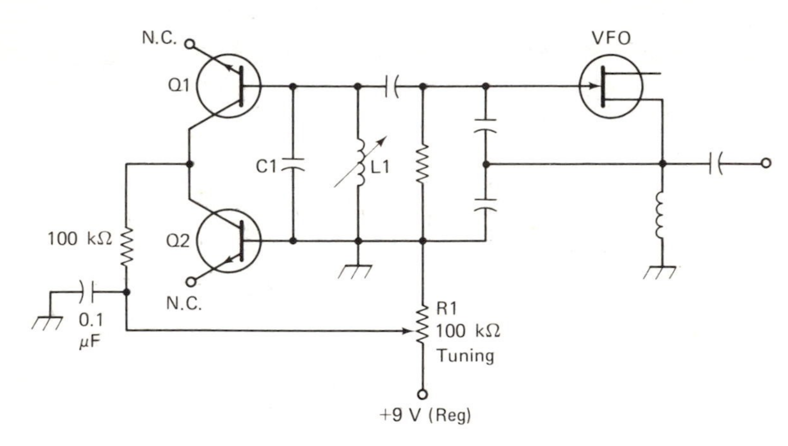

Figure 2.19 VFOs can employ electronic tuning by using varactor tuning elements. Bipolar transistors *Q*1 and *Q*2 are used in this circuit as varactor diodes.

symmetry. Some distortion will be apparent when only one diode is used.

The Motorola MV104 tuning diode, for one, contains back-to-back diodes. A typical circuit for this style of VVC diode is given in Fig. 2.20. It can be seen that *D*1 has a common cathode and separate anodes. The reverse bias is applied to the cathode. The remainder of the circuit is the familiar Colpitts oscillator we have discussed in this chapter.

Since we are interested at all times in the purity of the oscillator output spectrum, it is essential that we apply well-filtered dc bias to the tuning diode. The control voltage must be free of ripple, noise, and transient responses in order to prevent excessive bandwidth of the oscillator output frequency.

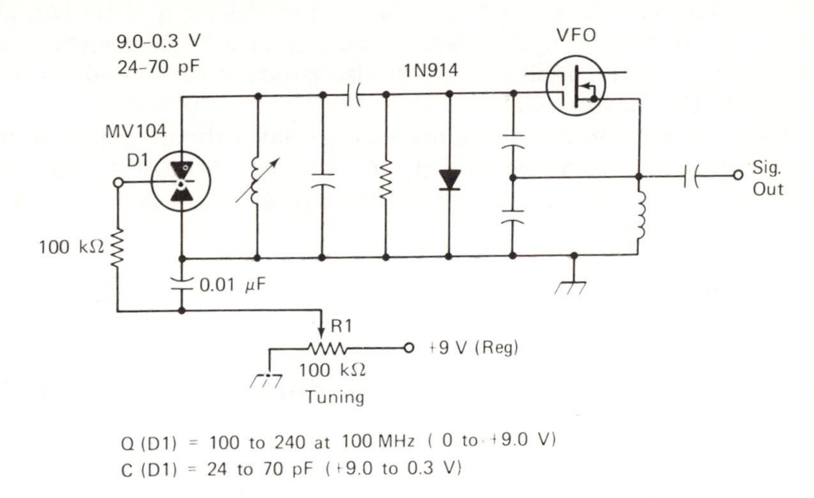

Figure 2.20 Circuit for a VFO that employs a conventional tuning diode (*D*1).

2.6 VFO LOAD ISOLATION

Changes in circuit reactance following the VFO will cause some change in the desired operating frequency. The severity of the "pulling" becomes worse as the operating frequency is increased. This phenomenon is often observed when crystal oscillators are used in the upper part of the high-frequency spectrum, particularly when a number of frequency multipliers are used to achieve VHF or UHF operation. When load changes are likely to occur because of the need to key or switch the load into which the VFO operates, it is to our advantage to use buffering or isolating stages after the oscillator. The number of isolation stages that follow the oscillator will depend upon the degree of load change seen by the VFO during normal operation. An example of the type of circuit that can cause problems is when a common VFO is used for transmitting and receiving in a transceiver. Another problem-causing circuit is a CW transmitter that has a VFO, but does not utilize the heterodyne technique of frequency generation. That is, the VFO operates on the transmitter output frequency directly. Severe chirp of the CW note can result when keying the stages that follow the VFO. The changes in reactance throughout the transmitter are reflected back to the oscillator during the keying period, thereby causing changes in oscillator frequency. Similarly, RF currents that are permitted to flow on the dc supply bus can cause shifts in oscillator frequency. Therefore, careful decoupling of the supply bus to each stage of the transmitter, plus adequate shielding of the oscillator section of the equipment, is almost mandatory toward the prevention of instability.

Self-oscillation in any of the stages that follow a VFO can also degrade the oscillator performance, causing erratic frequency instability. Such parasitic oscillations will also produce unwanted spurious output from the local oscillator.

We see from the foregoing discussion that a theoretically perfect *LC* type of oscillator can be ruined by a variety of outside causes. Let us examine some methods for resolving possible problems of the kind we addressed in this section.

2.6.1 RC Buffering

A simplistic but often effective method for isolating a VFO from its load is shown in Fig. 2.21. *Q*2 and *Q*3 form a direct-coupled buffer

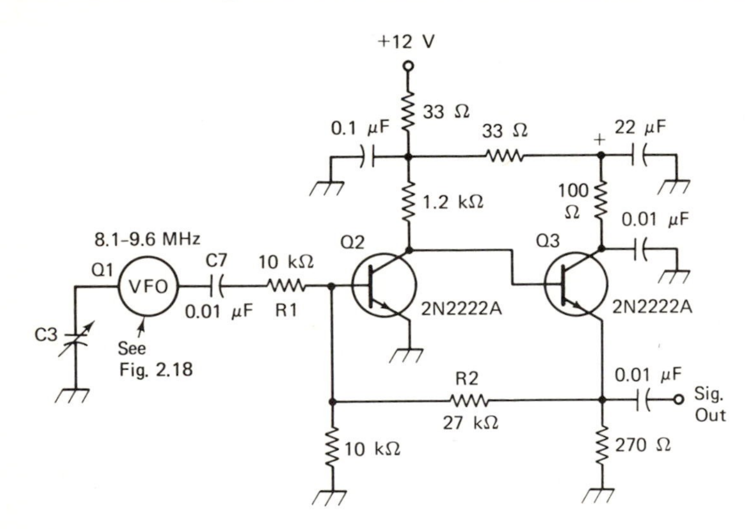

Figure 2.21 Details for an *RC*-coupled VFO buffer section (*Q*2 and *Q*3).

section, using dc feedback via *R*2. Capacitive coupling is used at the input and output of the buffer section. *R*1 reduces the coupling from the VFO to the input of *Q*2 and provides a high-impedance (10 kΩ) series component for the signal, an aid to isolating load changes. It should be noted, however, that if low oscillator noise is a design objective, the series resistance may contribute to the overall noise figure. This is because current flowing through a resistance generates noise, however miniscule it might be. For most applications the addition of *R*1 will not cause problems. The isolation characteristics of the *Q*2/*Q*3 transistor pair will be adequate for most circuits that do not reflect large changes in reactance. The trade-off with this circuit is low gain and the absence of the selective circuits for filtering the output waveform of the oscillator.

2.6.2 Isolation with Filtering

An improved type of buffer section, which includes amplification and filtering, is seen in Fig. 2.22. In this example we find a JFET being used at $Q2$ as a source-follower. The inherently high input impedance of the FET aids considerably in load isolation. $R1$ sets the Z_{in} characteristic at 100 kΩ. This affords significantly better isolation than the 10-kΩ resistor ($R1$) of Fig. 2.21. Furthermore, there is no series R in the signal path to contribute to the oscillator-chain noise output. $C7$ is a small-value capacitor (high X_c) to further reduce the effects of load changes.

The source of $Q2$ is broadly resonant over the oscillator frequency range. $RFC2$ forms the tuned circuit with approximately 10 pF of stray circuit capacitance. In this circuit a 32-μH choke satisfies the need.

A class A broadband amplifier provides approximately 10 dB of gain at $Q3$. Since the stage operates over a range of 1.5 MHz, extensive broadbanding need not be applied. Hence, there is no intentional feedback in the circuit. A 220-Ω resistor is used in parallel with $RFC3$ to broaden the response of the choke and to establish a collector characteristic of 220 Ω at $Q3$. A double-section pi network matches the $Q3$ output to a 50-Ω load, such as that of a doubly balanced diode-ring mixer. The network also filters out the harmonic currents present in the collector of $Q3$. We can see the effects of the harmonic filter in the spectrographs of Fig. 2.23. Although these responses were measured at 2.8 MHz, using a VFO-chain circuit identical to that of Fig. 2.22, the effects are the same. The display in Fig. 2.23(a) shows the output as taken at 50 Ω through a 4 : 1 broadband toroidal transformer, minus the harmonic filter. The second harmonic is 32 dB below fundamental output, and the third harmonic is down 18 dB. The spectral display of Fig. 2.23(b) shows the output with the two-section low-pass filter. The second harmonic is barely visible in the noise at the bottom of the display. The horizontal divisions are 1 MHz each, and the vertical divisions are 10 dB each. A Hewlett-Packard analyzer was employed during the tests.

An important design feature of the VFO is the low-impedance output at $L3$ and $C11$. Changes in reactance following the VFO chain of Fig. 2.22 will have minor effects on the operating frequency when the output of the chain has a low characteristic impedance. This is particularly helpful when the load following the chain is substantially higher in impedance, such as the base of a class A postamplifier stage. Normally, that type of load is on the order of 500 to 1500 Ω, depending on the biasing and style of amplifier used. The trade-off in this case would be

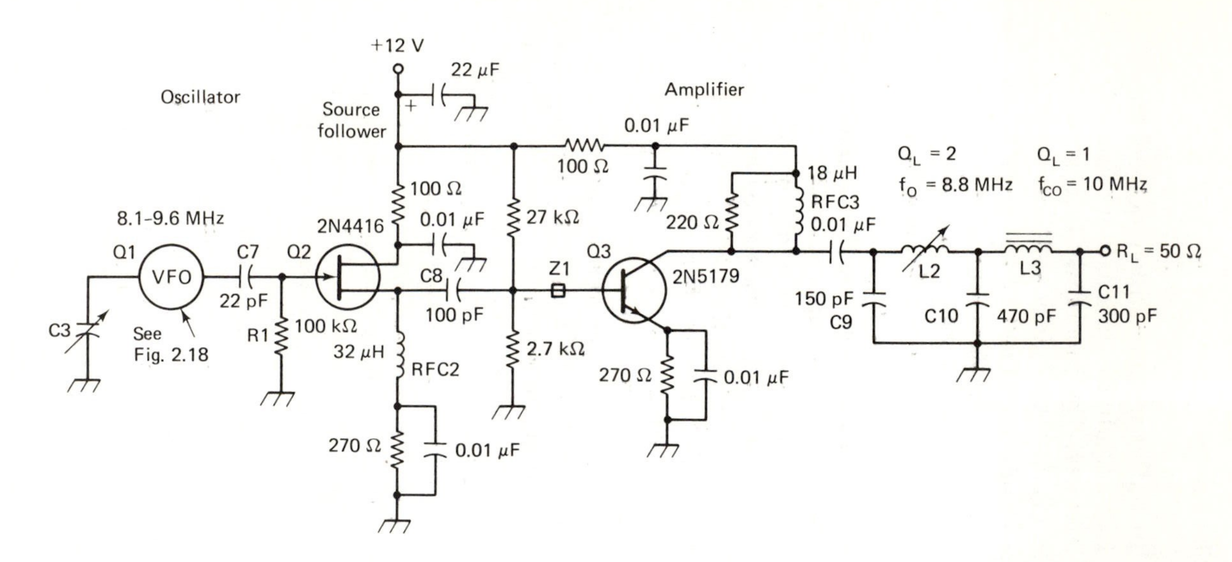

$X_{C8} \leqq 200\ \Omega$

$X_{C9} = 113\ \Omega$

$X_{C10} = 38\ \Omega$

$X_{C11} = 50\ \Omega$

$X_{L2\ (NOM)} = 105\ \Omega$

$X_{L3} = 50\ \Omega$

$X_{L\ (RFC2)} \approx 1810\ \Omega$

$X_{L\ (RFC3)} \geqq 880\ \Omega$

Figure 2.22 Practical VFO buffer circuit that contains an output filter to enhance spectral purity.

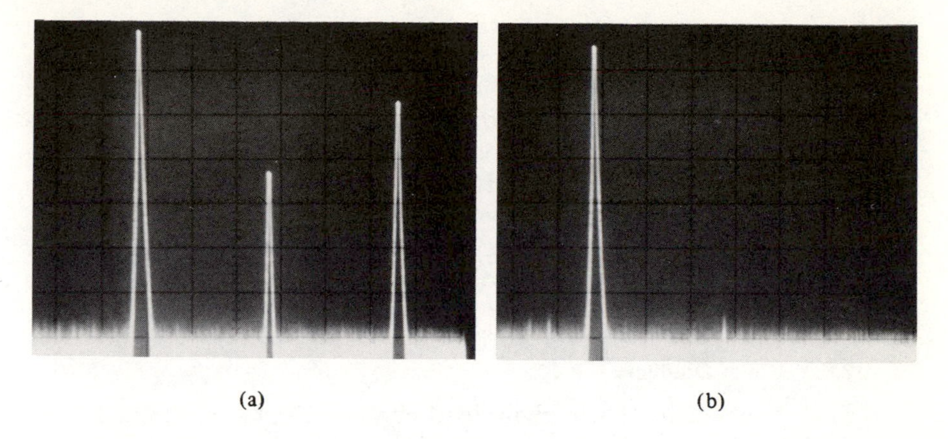

Figure 2.23 (a) VFOs can produce harmonic output that is objectionable, but (b) a simple harmonic filter at the output of a VFO chain can ensure a clean output spectrum.

degraded power transfer through intentional mismatch: the loss can be made up easily by adding another amplifier stage external to the VFO chain.

A parasitic suppressor, $Z1$, is used at the base of $Q3$ to prevent self-oscillation at VHF and UHF. The 2N5179 was chosen for the circuit because of its high F_T rating (1200 MHz). As a consequence, the amplifier can self-oscillate if the layout is not planned carefully. Double-sided PC board and short leads in that part of the circuit are suggested in the interest of good stability. $Z1$ is a miniature ferrite bead with a μ_i of 125. Some suggested reactances are listed in the diagram of Fig. 2.22

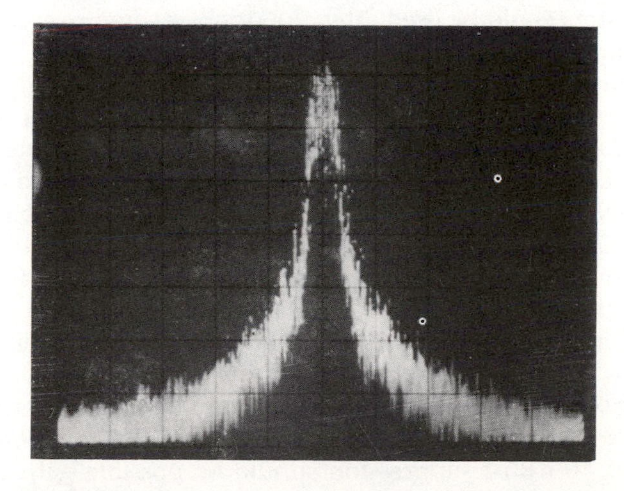

Figure 2.24 *LC* oscillators produce wide-band noise in their outputs, as seen in this spectrograph. The noise bandwidth can be reduced by using high-*Q* tuned circuits in the oscillator.

as an aid to designers. $Q1$ follows the circuit of Fig. 2.18. A typical noise-bandwidth profile for the circuit of Fig. 2.22, but at an operating frequency of 2.8 MHz, is shown in Fig. 2.24. The effective noise bandwidth is 0.1 kHz. Horizontal divisions are 2 kHz, and the scan time is 2 sec./division. The spectrograph was taken with a Hewlett-Packard analyzer. Oscillator noise bandwidth can be restricted through the use of high-Q tuned circuits. Hence, the lower the Q, the greater is the noise bandwidth. A low noise and spur level is essential for the local oscillator of a high-performance receiver and is a good design objective regardless of the application.

2.6.3 Use of Doublers

The effects of oscillator-chain load changes can be minimized additionally by utilizing an oscillator that operates one octave below the desired output frequency. For example, we might elect to design the oscillator for operation between 9 and 10 MHz to secure an output of 18 to 20 MHz. In a circuit of this type we would follow the VFO with a frequency doubler, and then amplify the 18- 20-MHz energy and filter it with suitable bandpass networks. The significance of this approach is that changes in load reactance at $2f$ have very little effect at f. Furthermore, it is much easier to achieve frequency stability at 9 MHz than it is at 18 MHz in an LC type of oscillator. We must be aware, however, of the possibility for generating spurious frequencies through the leakage of f at the output of the composite VFO chain and through any mixing of f and $2f$ in the post-doubler amplifiers. It is for this reason that bandpass tuned circuits or filters are important in this kind of circuit. Lowpass filters are ineffective in suppressing the f component. Although high-pass filters could be used after the doubler, they would not attenuate the harmonic energy of f and $2f$.

Push-push doublers are more suitable than single-ended doublers because they will not pass a significant amount of driving frequency if they are balanced well. Also, the push-push circuit approaches a straight class C amplifier in efficiency (70%), whereas a single-ended doubler has an efficiency on the order of 50%. In either event, the doubler should operate in class C to function effectively as a multiplier.

Two push-push doublers are shown schematically in Fig. 2.25. Additional treatment of these general circuits is given in Chapter 5. The circuit of Fig. 2.25(a) makes use of two JFETs in an efficient doubler. Dynamic balance is made possible by the adjustment of the variable source resistor, $R1$. This control and $C1$ are adjusted alternately until no further reduction in f can be observed on a scope at the output of $T2$. At least 40 dB of suppression should be possible with this circuit. Matched JFETs at $Q1$ and $Q2$ will aid dynamic balance. A matched, dual-JFET device, such as the Siliconix 2N5911, would serve well in this circuit.

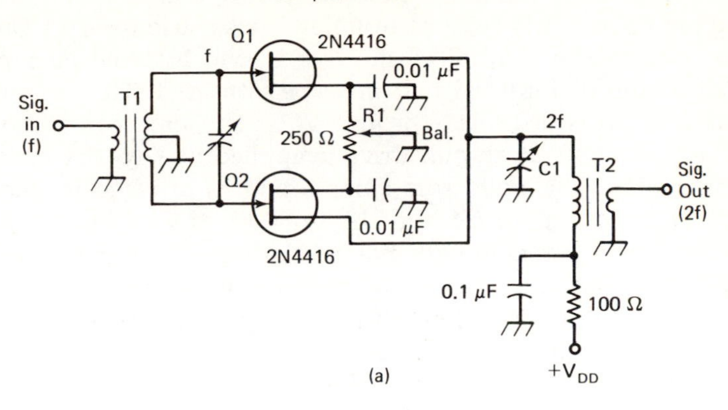

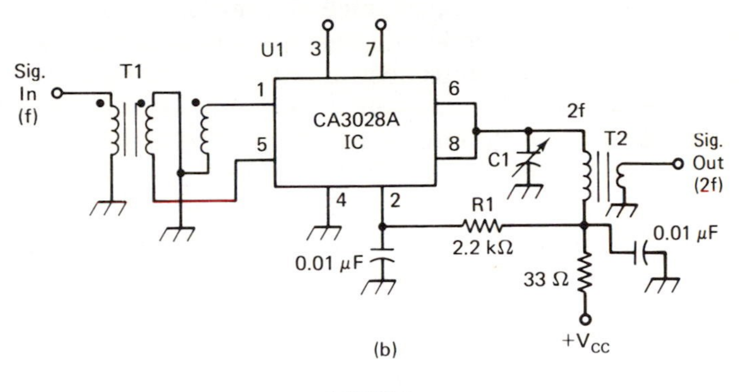

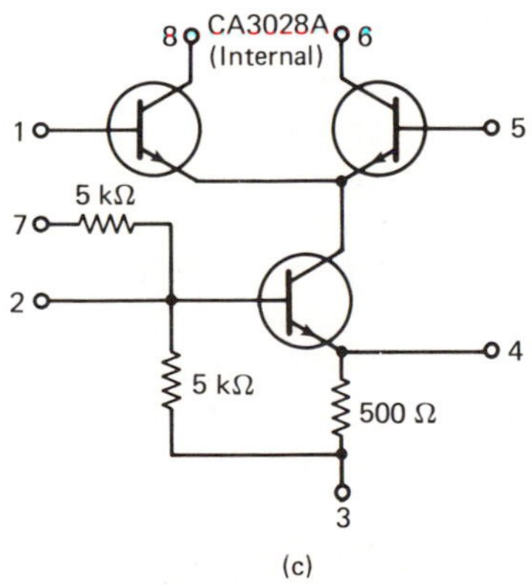

Figure 2.25 (a) Balanced push-push doubler using a pair of JFETs; (b) CA3028A IC is used for the same purpose; (c) circuit of a CA3028A IC.

Excellent dynamic balance can be obtained by using a differential-amplifier IC such as the RCA CA3028A. A circuit example is given in Fig. 2.25(b). Owing to internal bipolar transistors being fabricated at the same time on a common substrate, beta match and temperature tracking are good. This type of doubler is also discussed in Chapter 5.

To operate the CA3028A in class C, the internal current-source transistor of the differential pair must be saturated. This is achieved by application of forward bias through $R1$. The current-source emitter (pin 4) is grounded. Push-pull drive is applied at f to the bases of the differential pair. The collectors (pins 6 and 8) are tied in parallel and tuned to $2f$ by means of $C1$ and the primary of $T2$. Input transformer $T1$ can be tuned, or a broadband, trifilar-wound transformer can be used as shown. Suppression of f at the output of this circuit can be as great as 50 dB below $2f$. Other brands and types of differential ICs lend themselves well to this circuit arrangement.

Solid-state power doublers of the push-push variety are not recommended, because they tend to fall out of balance at they heat up, a result of changes in internal capacitance and resistance. A power doubler would be practical only if the equipment in which it was used operated continuously. In a continuous-duty application the doubler could be balanced after the transistors reached their normal operating temperature, thereby ensuring the required balance. VMOS power FETs, such as the Siliconix VN-66AK and VMP-4, would be suitable in a power push-push doubler, as they require substantially less excitation power than bipolar devices that could deliver comparable RF output. Additional coverage of doublers is provided in Chapter 5.

2.7 HETERODYNE FREQUENCY GENERATION

Direct frequency control, using a single VFO, is somewhat unpopular today in equipment that is required to cover a large part of the frequency spectrum. We can better appreciate this truism by acknowledging that in order to operate a transmitter from, say, 3 to 30 MHz, the VFO would require switched L and C components and would become progressively more unstable as the operating frequency was increased. The employment of a single VFO is satisfactory when a relatively small portion of a band of frequencies is to be accommodated by a transmitter or receiver (i.e., 3 to 5 MHz) in the lower part of the HF spectrum (10 MHz down to VLF).

When many segments of the HF spectrum need to be covered by a single piece of communications equipment, the heterodyne frequency-generation scheme becomes the practical solution, unless a frequency synthesizer is used as the local oscillator. Unfortunately, the present state of synthesizer art is such that a substantial amount of noise and spur energy will be present in the synthesizer output unless considerable sophistication and expense are included in the design effort. A mediocre or poor synthesizer, in terms of purity of output waveform, can ruin an otherwise excellent design. In the case of a high-performance receiver, excellent dynamic range can be negated by noise and spurious output from a synthesizer. Similarly, the spectral purity of a transmitter can be substandard because of in-band and out-of-band spurious products that originate in the synthesizer. This is by no means an indictment of frequency synthesis. Rather, our concern should be for the trade-off between expense and performance if a quality synthesizer cannot be justified in a particular design.

Still another consideration when choosing between a frequency synthesizer and an *LC* oscillator or heterodyne scheme is the efficiency of the overall equipment. Unless CMOS circuitry is utilized in a synthesizer, the power-supply requirements, in terms of current, can be prohibitive for certain equipment applications. The additon of LED-type digital frequency display will add substantially to the current taken from the power supply.

Finally, the amount of frequency resolution desired needs to be considered. Most synthesizers found in commercial transmitters and receivers for HF-band use have 100-Hz resolution, or increments, of tuning. At VHF and UHF this might be 5 kHz or more. Although a 100-Hz resolution is often adequate for SSB and CW operation, it tends to limit the general utility of the equipment and in some instances imposes a psychological handicap on the operator who has not been constricted previously by incremental tuning. The *LC* oscillator or heterodyne frequency generator provides freedom from the various conditions we have just discussed.

2.7.1 Heterodyne Circuit

A heterodyne frequency generator (HFG) consists of a tunable oscillator (or VXO) and a crystal-controlled oscillator. The frequency combination is chosen to yield a desired intermediate frequency (IF) by means of a mixer and suitable filtering. A hybrid diagram of 15.0- to 16.5-MHz HFG is shown in Fig. 2.26. In this example we are using the *difference* frequency as the IF, but the *sum* frequency would be suitable as an alternative with the proper choice of crystals at *Q*1.

Returning to Fig. 2.26, if we select *Y*1 by means of *S*1*C* and adjust the VFO for a frequency of 5.5 MHz, our difference frequency will be 15.0 MHz (20.5 minus 5.5). Conversely, if *Y*1 is kept in the circuit and the VFO is tuned to 5.0 MHz, the resultant IF will be 15.5 MHz at the output of *U*1.

The major advantage of this type of circuit is that a single tuning range is required for the VFO, negating the need for band switching within the VFO. This enables us to operate the VFO in some low part of the HF range, thereby making stability easy to achieve. Furthermore, if analog frequency readout is used, the VFO dial can be marked from 0

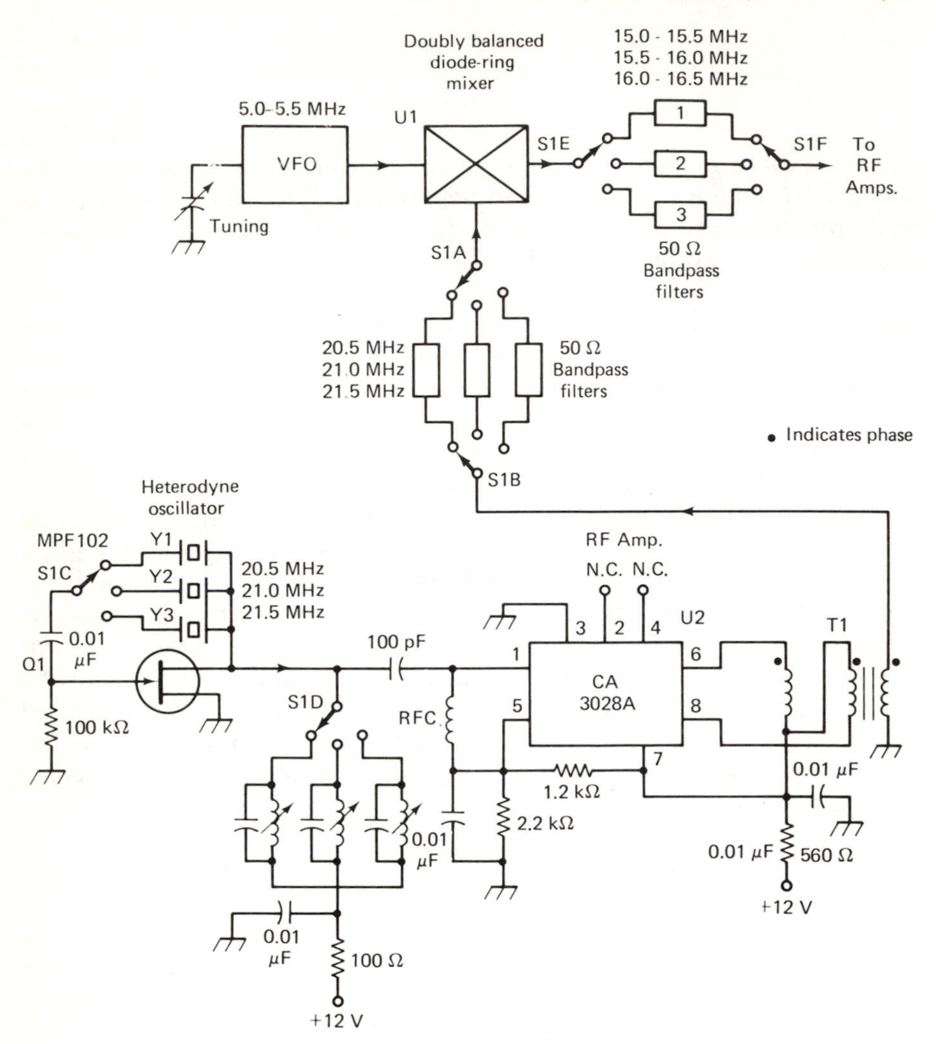

Figure 2.26 Hybrid circuit illustration of a basic heterodyne-frequency generator for use in multiband transmitters or receivers.

to 500 and will be accurate for any band covered by the HFG. Most analog dials have a coarse frequency dial (100-kHz increments), with fine resolution (1-kHz increments) being marked on the skirt of the main tuning dial.

Although mechanical switching is indicated in Fig. 2.26, diode switching could be used in place of *S1A* through *S1F*. Such an approach would be less costly and result in simplified layout and construction of the HFG module.

In our circuit example we find bandpass filters between RF amplifiers *U2* and the mixer. This will help to ensure that only the desired

LO (local oscillator) injection frequency reaches the diode-ring mixer. Similarly, bandpass filtering is incorporated at the mixer output to help reject the unwanted sum frequency mentioned earlier. The filters will also suppress the heterodyne oscillator frequencies and that of the VFO. A further aid to purity of waveform at the output of the mixer is the use of a doubly balanced type of mixer (*U*1). An inherent feature of the doubly balanced mixer (DBM) is cancellation of the energy applied to one port, as measured at either of the remaining ports.

A single-ended active mixer, such as a dual-gate MOSFET, would serve in place of *U*1, but suppression of unwanted frequencies at the mixer output would be much more difficult than with the method illustrated in Fig. 2.26. As still another aid to clean output from the mixer, harmonic filtering can be applied between the VFO and *U*1. Normally, this takes the form of a simple half-wave low-pass filter with a loaded *Q* of 1. The noise output from an HFG of this type will be quite low if high-*Q* resonators are used in the VFO and if the mixer injection levels are proper. This is discussed in Section 6.1. The circuit of Fig. 2.26 would be suitable for use in a transmitter or to inject a mixer in a superheterodyne receiver. The HFG subject is also discussed in Chapter 5.

2.8 AUDIO OSCILLATORS

Audio sine-wave oscillators are used in some types of communications equipment, which justifies inclusion of some basic information about them in this chapter. They are used in CW transmitters as side-tone monitors. Another application would be as audio generators for AM transmitters, such as an aircraft radio beacon transmitter, in which a modulated continuous wave (MCW) identifier is used to identify the carrier of a particular airport. For all these applications it is essential that the audio waveform be as pure as possible. This rules out the use of relaxation oscillators of the multivibrator, unijunction transistor, and neon-lamp types.

2.8.1 Phase-Shift Oscillator

The principle of oscillation remains unchanged whether or not the oscillator operates at radio or audio frequencies. Fundamentally, we are still dealing with sine waves and amplifiers to which feedback voltage is applied. The circuit of Fig. 2.27 is an *RC* resonant feedback oscillator, which is commonly referred to as a *phase-shift oscillator*. However, audio oscillators can be designed to operate with *LC* components also. An example of this is seen in the Touchtone pads used by Bell Telephone and other phone companies. The tone generators use high-*Q* pot-core inductors in an *LC* type of circuit.

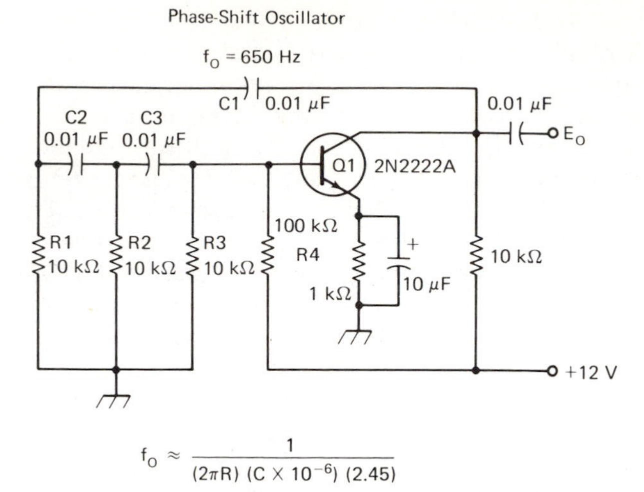

$$f_O \approx \frac{1}{(2\pi R)(C \times 10^{-6})(2.45)}$$

Figure 2.27 Basic phase-shift audio oscillator.

The *RC* network in Fig. 2.27 contains three sections. Each section ($C1/R1$, $C2/R2$, and $C3/R3$) helps to cause a phase shift of 60° at f_o, the operating frequency. The capacitive reactance of the network increases or decreases at frequencies above and below f_o, thereby permitting the required 180° phase shift only at f_o. This ensures that the oscillator output is stationary at the design frequency. Oscillators of this kind can be made to cover a specified frequency range by making the *R* or *C* elements variable by means of ganged capacitors or potentiometers.

Feedback losses can be minimized by using additional *RC* sections of the same value as the first three. However, the circuit of Fig. 2.27 should be ample for most applications. As in the case with high-frequency *LC* oscillators, light coupling should be used at the *RC*-oscillator output to minimize frequency pulling when the load changes during keying.

2.8.2 Wien-Bridge Oscillator

A standard audio oscillator for the past several decades has been the Wien-bridge type. It also uses *RC* components and provides a clean sine wave at the output. A typical circuit in which an operational amplifier (op-amp) is used can be seen in Fig. 2.28. $R1/C1$ and $R2/C2$ provide the feedback path around $U1$. *I1* and $R3$ supply negative feedback for the circuit. At the design frequency (f_o), the phase shift is zero. Positive feedback causes oscillation at f_o under the conditions indicated in the equation of Fig. 2.28. In this case *R* is in ohms, *f* is in hertz, *C* is

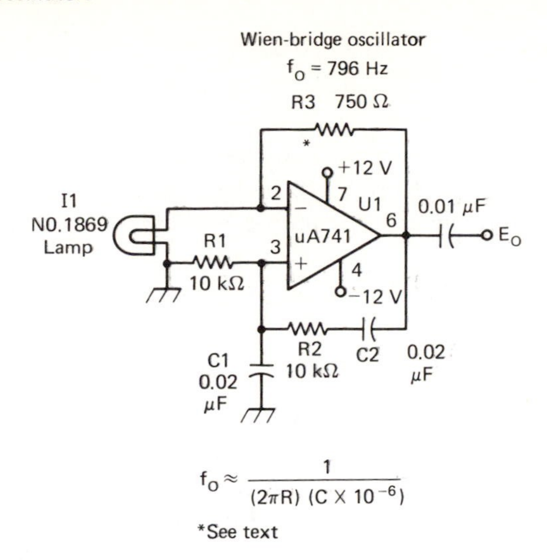

Figure 2.28 Good waveform purity and frequency stability can be obtained from this Wien-bridge oscillator.

in microfarads and π = 3.14. The same units are used in the equation contained in Fig. 2.27.

An op-amp is specified in this circuit, but discrete transistors or vacuum tubes could be used as an alternative. The gain bandwidth of the op-amp chosen should be at least ten times greater than f_o to assure reliable operation and optimum signal output.

The nonlinear characteristics of the lamp (*I*1) are necessary in the negative feedback divider to maintain the negative feedback at the same level as the positive feedback (necessary to oscillation). Other devices can be used for this part of the feedback divider, such as thermistors or Zener diodes, but the lamp is preferred by many designers because of its simplicity and low cost.

By keeping the positive and negative feedback voltages equal, the bridge can be considered balanced, and the differential input to *U*1 will be low. Too much positive feedback will cause the op-amp to saturate, thereby stopping oscillation. Conversely, too much negative feedback will cause the oscillations to languish until the circuit becomes inoperative.

The negative feedback can be adjusted precisely by replacing *R*3 with a fixed-value resistor and a variable resistor. A 500-Ω potentiometer and a 470-Ω resistor in series will suffice for this circuit. Variable-frequency operation is practical with this circuit by making the *R* or *C* elements of the positive-feedback network variable. As with the circuit of Fig. 2.27, the variable elements must be ganged and made to track accurately to keep the feedback ratio fixed.

BIBLIOGRAPHY

Papers

DeMaw, D., "Building a Simple Two-Band VFO," *QST Magazine*, June 1970.

—— "Some Practical Aspects of VXO Design," *QST Magazine*, May 1972.

—— "VFO Design Techniques for Improved Stability," *Ham Radio Magazine*, June 1976.

Books

Cowles, L. *A Source Book of Modern Transistor Circuits*, Chapter 14, Prentice-Hall, Inc., Englewood Cliffs, N.J., 1976.

DeMaw, D., *Practical Communications Data for Engineers and Technicians*, Chapter 3, Howard W. Sams & Co., Inc., Indianapolis, Ind., 1978.

Hayward W., and D. DeMaw, *Solid State Design for the Radio Amateur* (various sections on oscillator design), ARRL, Inc., Newington, Conn., 1977.

Krauss, H., C. Bostian, and F. Raab, *Solid State Radio Engineering*, Chapter 5, John Wiley & Sons, New York, 1980.

Lenk, J., *Handbook of Simplified Solid State Circuit Design*, Chapter 5, Prentice-Hall, Inc., Englewood Cliffs, N.J., 1971.

3

SMALL-SIGNAL RF AMPLIFIERS

Small-signal RF amplifiers function in a like manner, irrespective of the operating frequency. The notable exception to this rule occurs when we consider the operating mode: class A, AB, B, or C. Therefore, the amplifier stage might find application in a receiver front end, an IF amplifier strip, LO-chain (local oscillator) amplifier, or the low-level section of a transmitter. The operating mode will depend largely upon the available excitation from the preceding stage and whether or not amplifier linearity is a criterion.

RF amplifiers may be configured as common-emitter, common-base or common-collector stages. Similarly, the equivalent circuits for FETs or vacuum tubes are entirely suitable in a design we might choose to develop.

The dynamic range of the amplifier is still another important consideration. This means that the device should be chosen carefully to ensure that it will handle the applied signal energy without causing undue distortion somewhere below the level at which saturation will take place. Similarly, the transistor (or IC, if one is used) must be permitted to operate below its maximum safe dissipation power level.

Finally, designers need to consider the various conditions under which amplifier instability can occur. Steps can be taken early in the design and layout exercise to prevent self-oscillations at, near, or far from the operating frequency. A special section in this book will address the subject of amplifier stability. The cause of instability is unwanted *feedback*, the component that is required for oscillator circuits, as outlined in Chapter 2.

3.1. AMPLIFIER CONFIGURATIONS

Most designers prefer the common-emitter (grounded emitter) or common-source amplifier arrangement when working with bipolar transistors or FETs. This is because maximum gain can be realized in this configuration, and the input impedance of the amplifier is higher than that of the common-base or common-gate amplifier. Typically, the input characteristic of a class A small-signal bipolar-transistor amplifier is on the order of 500 to 1500 Ω, depending on the biasing and choice of component values external to the amplifier input terminal. Conversely, the common-base amplifier may present an input impedance as low as 50 Ω, contingent upon the transconductance of the device. A possible advantage is, however, a relatively high output impedance with this style of amplifier.

3.1.1 Common-Base Amplifiers

Examination of Fig. 3.1 will indicate that for a common-base amplifier the signal is applied between the emitter and base. The load is

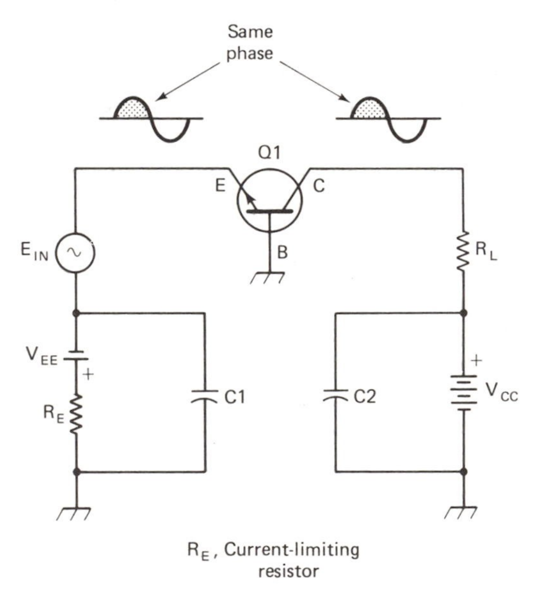

Figure 3.1 Arrangement for a common-base amplifier showing phase relationship of input and output waveforms.

connected between collector and base. Supply voltages V_{EE} and V_{CC} determine the correct biasing of the device. $C1$ and $C2$ serve as bypass capacitors for the power supplies, thereby ensuring that the full ac signal is developed across R_L. Variations in emitter current result from application of the input signal. These variations cause the collector current to vary.

The current gain of the amplifier seen in Fig. 3.1 is the ratio of the ac output current to the ac input current. This gain is symbolized by A_I. In our circuit the output current represents the collector current (I_C), and the emitter current (I_e) represents the transistor input current. Therefore, current gain can be expressed by

$$A_I = \frac{I_C}{I_e}$$

From this we can see the current gain of a common-base amplifier is roughly equal to alpha (α), which means it will always be less than 1. In other words, there will always be a slight current loss in this type of amplifier. Conversely, the amplifier can provide voltage and power gain.

Voltage gain (A_V) is the ratio of the output voltage to the input voltage. For our common-base circuit of Fig. 3.1 the output voltage (V_C) is the collector voltage, and the input voltage (V_E) is the emitter voltage. From this we obtain

$$A_V = \frac{V_C}{V_E}$$

The voltage gain of our circuit is also equivalent to the product of α and the ratio of the load resistance to the input resistance of the amplifier. Input resistance is defined as that which appears between the transmitter emitter and base terminals. The foregoing relationship is expressed by

$$A_V = \frac{\alpha R_L}{R_{\text{in}}}$$

Power gain is defined as the ratio of the output power to the input power (G_P). For a common-base amplifier we are referring to the collector and emitter powers. Hence

$$G_p = \frac{P_C}{P_E}$$

where P_C is the collector power and P_E is the emitter power. We can see that G_P is equal also to the product of the current and voltage gain. This is expressed as

$$G_P = \frac{\alpha^2 R_L}{R_{in}}$$

We might sum up this discussion by saying that G_P is equal to α^2 times the resistance ratio of the common-base amplifier.

The waveforms at the top of Fig. 3.1 indicate the input-output phase relationship of the common-base amplifier. It can be seen that there is no phase reversal with this circuit or the equivalent FET amplifier of Fig. 3.2. Generally, the rules for amplifier gain, specified for Fig. 3.1, can be applied to the FET amplifier of Fig. 3.2. However, the de-

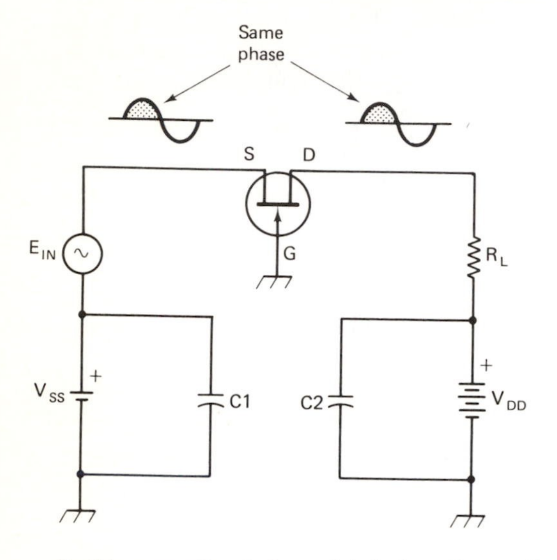

Figure 3.2 Common-gate JFET amplifier circuit showing input and output signal phase as being the same.

finitive method for obtaining the voltage-gain characteristic of a common-gate amplifier requires knowledge of the device g_{fs} (forward transconductance), as well as the terms expressed for bipolar-transistor amplifiers. A_V is expressed specifically by

$$A_V = \frac{(g_{fs}\, r_{os} + 1)\, R_L}{(g_{fs}\, r_{os} + 1)\, R_{\text{in}} + r_{os} + R_L}$$

where g_{fs} is the transconductance in microsiemens (μS), r_{os} is the FET common-source output resistance in ohms, R_{in} is the resistance of the input-signal source, and R_L is the output resistance inclusive of the load. Therefore, if we had a transistor with a g_{fs} of, say, 4000 S, an r_{os} (typical) of 7800 Ω, an R_{in} of 600 Ω, and an R_L of 3000 Ω, the common-gate voltage gain would be 5. If R_{in} were reduced to 200 Ω, the A_V would increase to 15, which clearly illustrates the relationship between the input and output load conditions of the common-gate FET amplifier. The characteristic input impedance of a common-gate FET amplifier is determined by

$$Z_{\text{in}} = \frac{10^6}{g_{fs}} \quad \Omega$$

R_{in} is, therefore, this value in parallel with the impedance of the signal source. Thus, if the g_{fs} were 4000 S, the characteristic Z_{in} of the source terminal would be 250 Ω. With a signal source of 600 Ω, the resultant R_{in} would equal 176.5 Ω. This rule holds true for the source-follower configuration, which is described later as the common-drain amplifier.

3.1.2 Common-Emitter Amplifier

In the common-emitter amplifier, the emitter of the transistor is common to the input and output circuits, as illustrated in Fig. 3.3. In-

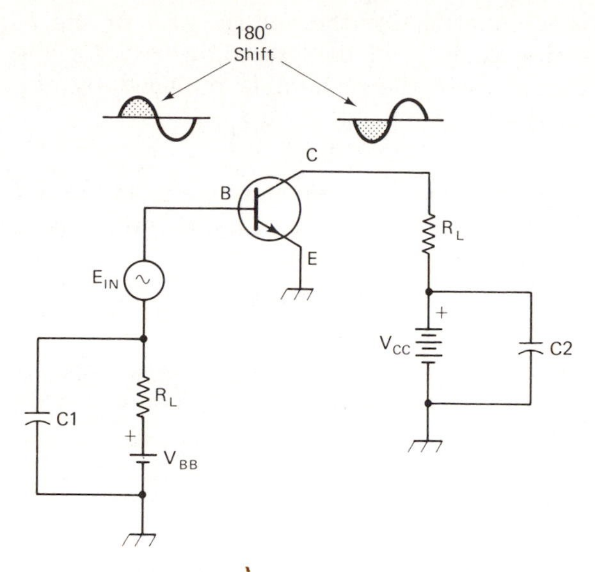

Figure 3.3 Common-emitter amplifier showing 180° phase shift of the input signal.

put energy is supplied across the base emitter (as in Fig. 3.1), but the output is taken across the emitter and collector. The base and collector are biased positive with respect to ground, thereby eliminating the need for separate batteries of power supplies.

Current gain from the common-emitter amplifier is a result of the input signal being supplied to the base, where it adds to or subtracts from the relatively minuscule base current of the transistor. With the base current being a specified percentage of the emitter current, the small changes in base current will cause proportionally larger variations in emitter current. As a result of this action, there will be a change in the collector current. The net effect of this is that we are using a small change in base current to effect a large change in collector current. Practically, then, the base current (I_B) is the input current of the amplifier, and the collector current (I_C) is the output current. From this evolves

$$A_I = \frac{I_C}{I_B}$$

where A_I is the current gain of the amplifier. This gain is also the equivalent of $\alpha/1 - \alpha$. The quotient is termed the current-amplification factor and is symbolized by beta (β). Hence

$$A_I = \beta = \frac{\alpha}{1 - \alpha}$$

For most small-signal bipolar transistors the range of β extends from approximately 20 to 200 or greater.

The basic equation for the voltage gain of the common-emitter amplifier is the same as in the preceding text for the common-base amplifier, except that the α term is replaced by the β. Therefore,

$$A_V = \frac{\beta R_L}{R_{in}}$$

Thus, if the input resistance for a common-emitter amplifier was 800 Ω and the output resistance (inclusive of the external load resistance) was 8000 Ω, the A_V could be determined by

$$A_V = \frac{50 \times 8000}{800} = 500$$

assuming in this example that $\beta = 50$. The typical collector resistance of a small-signal amplifier of the type depicted in Fig. 3.3 is on the order of 20,000 Ω. But this value is usually shunted with some lower value of load resistance with 10,000 Ω being rather typical. Therefore, the actual output resistance taken into account when we determine the A_V is substantially less than 20,000 Ω.

It can be seen that there is a 180° phase shift in the input signal by the time it appears at the collector. This is shown at the top of Fig. 3.3. A like situation prevails in the case of a common-source FET amplifier, as shown in Fig. 3.4. The voltage gain for a common-source amplifier (without feedback) can be determined by

$$A_V = \frac{g_{fs}\, r_{os}\, R_L}{r_{os} + R_L}$$

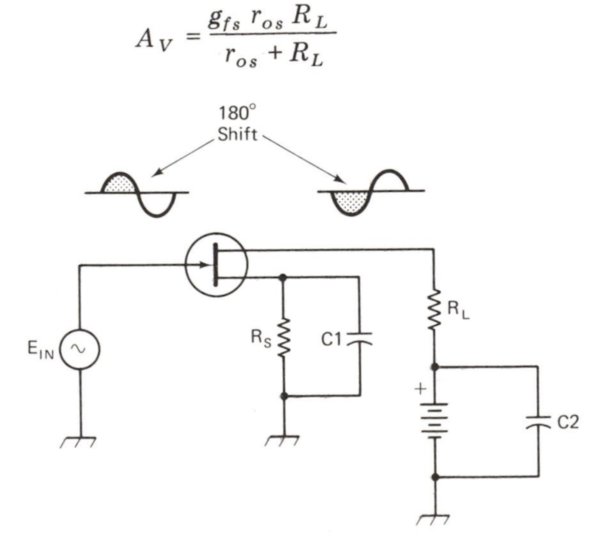

Figure 3.4 Common-source FET amplifier with 180° phase shift.

where the terms of the equation are the same as those for the common-gate amplifier. A negative feedback voltage that is proportional to the output current will be produced if the source contains an unbypassed resistor. The voltage gain with this degenerative feedback condition is expressed by

$$A_V = \frac{g_{fs}\, r_{os}\, R_L}{r_{os} + (g_{fs}\, r_{os} + 1)\, R_s + R_L}$$

where R_s is the total unbypassed source resistance in series with the source terminal of the FET. The common-source output impedance (Z_o) with this type of feedback is elevated by the unbypassed source resistor, and is determined by

$$Z_o = r_{os} + (g_{fs}\, r_{os} + 1)\, R_s$$

where Z_o is in ohms. The remaining terms of the equation were defined earlier in this section.

Let us return once more to the common-emitter bipolar transistor and consider its frequency characteristic as opposed to α for the common-gate amplifier. The common expression for any transistor amplifier is the *cutoff frequency*. As we learned earlier, this is expressed as α for the common-base amplifier. For a common-emitter stage it is called f_T. This is the frequency at which the device gain is unity (1). This is not to be confused with β for the common-emitter amplifier, which like α is the frequency at which the gain drops to 0.707 times its 1-kHz value. The f_T is also known as the *gain-bandwidth product*, the point at which the common-emitter forward-current transfer ratio (β) is unity. These characteristics are useful to us in selecting the best transistor or circuit arrangement for a specific application. As a viable rule of thumb, many designers choose a transistor that has a cutoff frequency of five to ten times greater than the intended operating frequency. This is especially significant when designing broadband, feedback types of amplifiers in order to ensure a reasonably flat response well into the upper part of the spectrum the amplifier must accommodate. The notable exception to this rule is in the case of a transistor oscillator. The device can be made to oscillate at its f_T or higher because of the positive feedback that is applied to the circuit.

3.1.3 Common-Collector Amplifier

We will examine next the common-collector amplifier, which is also called a source-follower in the case of an FET or an emitter-follower when the amplifier utilizes a bipolar transistor. In this configuration the collector (or drain of the FET) is grounded for alternating currents, as shown in Figs. 3.5 and 3.6. The signal is applied to the base-collector circuit (or gate-drain circuit) and is taken from the emitter-collector terminals. With this arrangement the input impedance of the amplifier is high and the output impedance is low. This makes the circuit useful for impedance transformations from high to low. The trade-off is that the voltage gain is less than unity (1) and the power gain is less than that obtained with the common-base or common-emitter circuits.

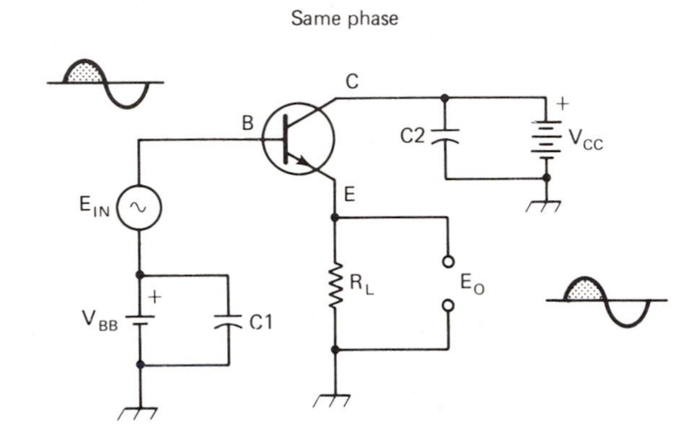

Figure 3.5 Common-collector amplifier. There is no phase shift.

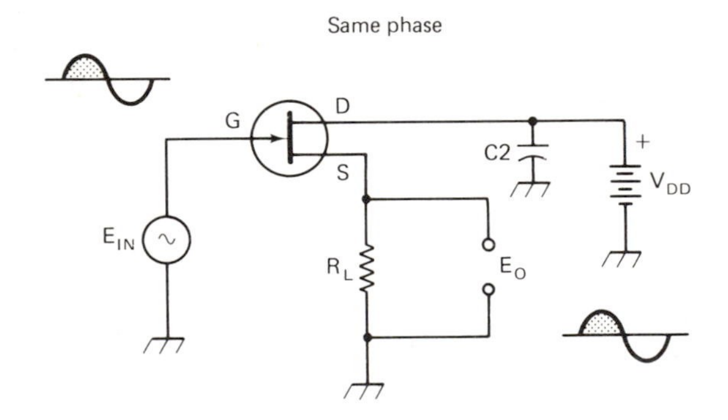

Figure 3.6 Common-drain amplifier showing no reversal of signal phase.

Z_{in}, the input impedance, is established by the characteristic impedance of the signal source, the natural resistance of the device input, and the value of bias resistors used at the input. The parallel combination of these components establishes R_{in}. An FET has an input characteristic of 1 megohm (MΩ) or greater, so the actual Z_{in} is usually set by the gate resistor in parallel with the impedance of the signal source. When dealing with a class A bipolar transistor, the typical input impedance (exclusive of external influencing factors) is between 500 and 1500 Ω.

As we learned earlier in this chapter, the output impedance of an emitter-follower or source-follower is similar to the input impedance of a common-base or common-gate amplifier. If the g_m (transconductance) of a bipolar transistor is known, Z_o can be determined from

$$Z_o = \frac{10^6}{g_{m\ s}} = \Omega$$

and g_m is extracted from

$$g_m = 10^3 \left(\frac{\Delta I_{C\ (\text{mA})}}{\Delta V_{eb}} \right) \text{S}$$

where Δ is the change in collector current for a change in emitter-base voltage. As is true of the previous two amplifier configurations, the actual Z_o will be influenced by the emitter or source resistor and the impedance of the external load. Furthermore, this type of amplifier does not provide a phase shift between the input and output signal, as shown in Figs. 3.5 and 3.6.

The most common of the three transistor configurations we have discussed is the common-emitter or common-source circuits. In summary, we can state the usefulness of the common-base amplifier as that of an impedance-matching active circuit. It is also beneficial when we desire the highest possible cutoff frequency for a given transistor. The common-emitter stage is widely used because it has the highest power gain of the three circuits, exhibits a fairly high input impedance, and lends itself best to cascading of stages. A common-collector amplifier is best suited to use as an active impedance-matching device for high-source and low-load conditions. Table 3.1 lists the general characteristics of the three modes of operation for bipolar transistors.

TABLE 3.1 Comparative Characteristics of Transistors

Characteristic	Common Base	Common Emitter	Common Collector
Voltage gain	Up to 60 dB	Up to 55 dB	Less than unity
Current gain	Less than unity	Up to 50	Up to 50
Power gain	Up to 30 dB	Up to 40 dB	Up to 16 dB
Phase reversal	none	180°	none
Frequency response	Highest	Lowest	Low
Input resistance	50 to 300 Ω	500 to 1500 Ω	35 to 45 kΩ
Output resistance	Up to 500 kΩ	Up to 75 kΩ	Less than 1 kΩ

The above characteristics are approximate in range. Actual parameters are dependent upon operating voltages and currents, and actual resistances will be influenced also by external load values and signal-source values.

Some representative RF types of circuits for bipolar and field-effect transistors are given in Fig. 3.7. The common-collector amplifier has not been included, since it closely follows the rules for the common-base circuit. Figure 3.7(a) shows a common-emitter RF amplifier that has its base tapped down on $L2$ to effect an impedance match and to prevent excessive loading of the resonator. The actual input impedance of the stage will be dependent in part on the value chosen for bias resistor $R1$. If the resistance is low, it will become significant as a parallel component to the characteristic input resistance of the transistor. If it is, say, kΩ or greater in value, its effect will be minor. The collector of $Q1$ is tapped down partway on $L3$ to preserve the resonator Q by virtue of lighter loading. The technique of tapping the transistor elements down on the tuned circuits will result in a loss of stage gain unless the tap points are selected to provide a close match between the transistor input and the signal source, and between the transistor collector and the output load. $C1$ should have a reactance of 10 Ω or less. The same is true of $C2$ and $C3$, unless $C3$ is used to control degenerative feedback, in which case its reactance will be made progressively higher as the feedback voltage is elevated. Figure 3.7(c) shows a JFET equivalent of the circuit of Figure 3.7(a). $R1$ is used to set the bias of the amplifier.

A common-base RF amplifier is seen in Fig. 3.7(b), with its FET equivalent shown at D. $R1$ is the normal bias resistor used in the common-emitter circuit. $C1$, $C2$, and $C3$ should have a reactance of less than 10 Ω in order to be effective as bypass capacitors.

3.1.4 Dual-Gate MOSFETs

Our discussion thus far has centered around bipolar transistors and JFETs. Essentially, the same rules apply to dual-gate MOSFETs as to the devices we have already treated. The primary difference is in the biasing of the control gate, or gate 2. A circuit example of a typical dual-gate MOSFET amplifier is provided in Fig. 3.8. $Q1$ can be one of the many dual-gate FETs available on the market, such as the RCA 3N200, RCA 40673, or Texas Instruments 3N211. The device transconductance (g_{fs}), as a function of the gate 1 to source voltage, can be controlled by the amount of bias applied to control gate 2. This is illustrated in Fig. 3.9 by a set of curves that demonstrate the transconductance versus the gate 2 bias. The maximum g_{fs} condition occurs near the point where the gate 1 to source voltage is –0.45 and the gate 2 voltage is +4. Transconductance under these conditions will be approximately 10,500 S with a drain current of 10 milliamperes (mA) and V_{DS} of 15. A source resistor of 270 Ω is a suitable choice for establishing the desired gate 1 to source voltage for most dual-gate MOSFET amplifiers. Complete cutoff of the transconductance will not occur unless

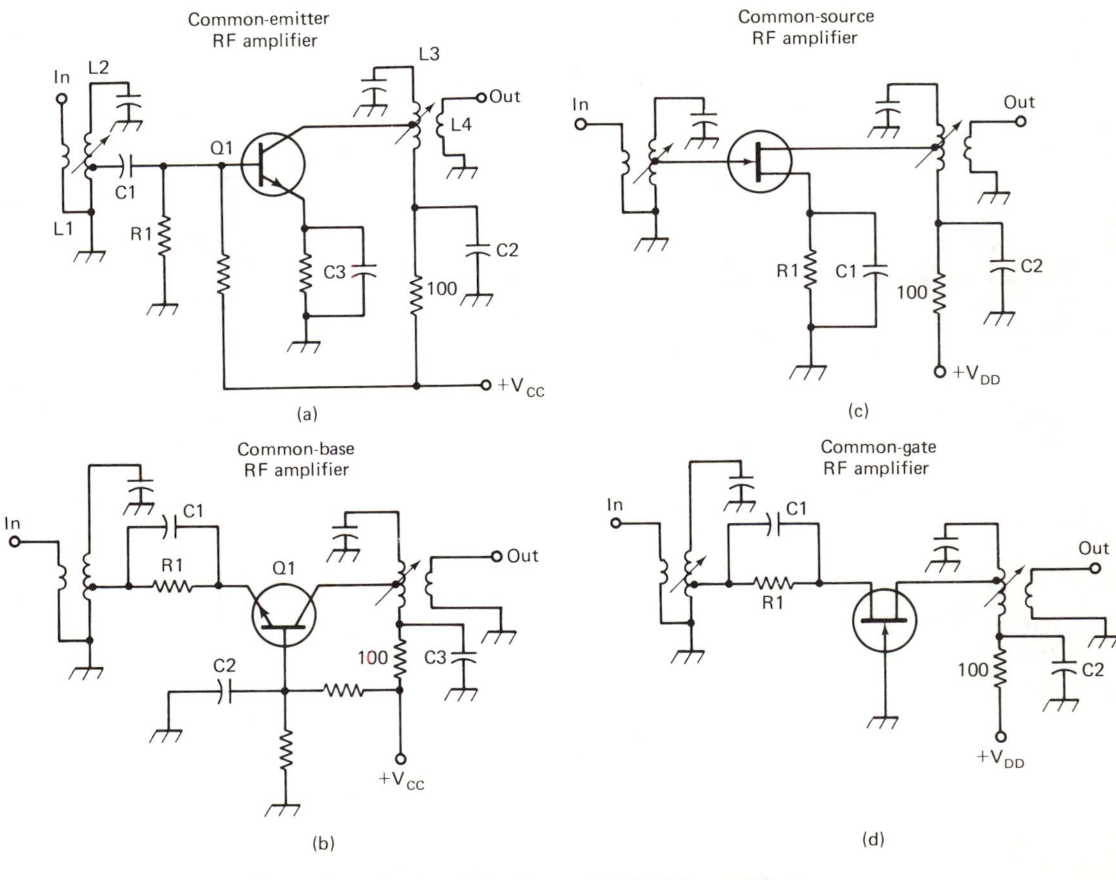

Figure 3.7 Practical bipolar and FET RF amplifiers.

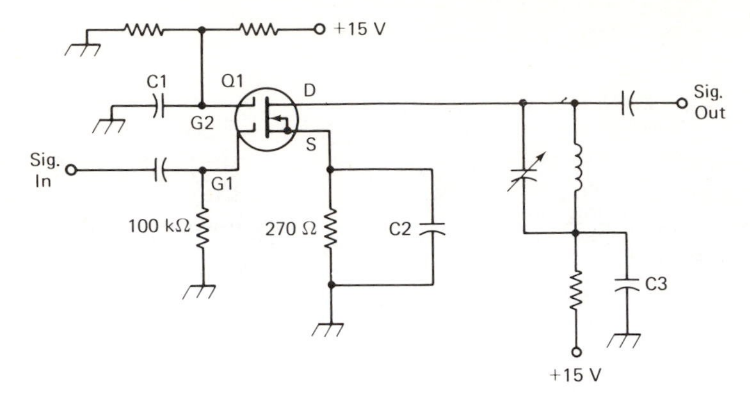

Figure 3.8 A dual-gate MOSFET works well as a small-signal RF amplifier.

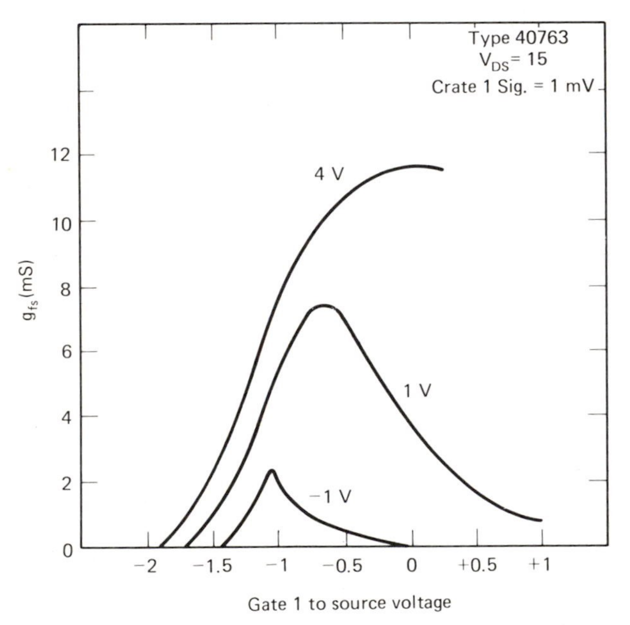

Figure 3.9 Curves that illustrate the effect of gate 2 voltage versus transconductance for a dual-gate MOSFET amplifier.

gate 2 is biased to approximately -2 V. Therefore, if we choose to apply AGC to a dual-gate MOSFET stage it is desirable to raise the source approximately 2.5 V above ground and reference gate 1 to the source. Then, when zero volts occurs at gate 2, we will have the equivalent of driving gate 2 in the negative direction by 2 V. Insertion of a Zener diode of appropriate value in the source return will achieve this. Similarly, an LED and a silicon diode can be connected in series and used in place of the Zener diode, as the LED will conduct at 1.5 V to establish part of the desired reference voltage. An additional 0.7 V will be added by the silicon diode, yielding a reference voltage of approximately 2.2 V.

$C1$, $C2$, and $C3$ of Fig. 3.8 should have low reactance in order to fulfill the bypassing requirements at those circuit points. $R1$ sets the input impedance of $Q1$ at 100,000 Ω, since the characteristic input resistance of the FET is in excess of 1 MΩ.

3.2 NOISE FIGURE

Noise is present in all types of amplifiers. Part of the overall noise component is generated within the transistor, while the remainder of it originates in the components associated with the amplifier, notably the input port of the stage. In the case of a receiver RF amplifier, noise not only comes from the aforementioned sources; it also arrives via the antenna. The latter is commonly referred to as *antenna noise*, and it is a term used frequently by designers or receivers for VHF and UHF operation. Below approximately 40 MHz, much of the antenna noise comes from the man-made sources, which combine with normal atmospheric noise.

If we could devise an *ideal* amplifier stage, there would be no noise generated internally or externally. But, as long as current is permitted to flow, however minuscule, noise will be generated. Therefore, we must strive to minimize the noise power with respect to the signal power in order to ensure the best *noise figure* possible. The net effect of excessive noise is the impairment of weak-signal reception, since the noise can be greater in magnitude than a signal that would pass unimpaired in an ideal amplifier or receiver. The reference used when determining *noise figure* is, therefore, the ideal amplifier or receiver.

There is considerable misunderstanding about the meaning of receiver *sensitivity*. It is not the ability of a receiver to produce large amounts of audio output power for a given level of small-signal input power. Rather, the sensitivity of a receiver can be defined as the input signal necessary to provide a specified ratio (in decibels) of signal-plus-noise output over the noise output of the receiver.

3.2.1 Character and Definition of Noise

Under the conditions of *thermal agitation* (random motion of electrons), noise will be proportional to the prevalent temperature, irrespective of the operating frequency if the bandwidth and input impedance of the receiver are held constant. Noise that is related to the foregoing condition is usually expressed as the *equivalent noise resistance*. Specifically, we are referring to the value of resistance (located at the input port of a stage) that will cause an output-port noise that equals the noise of the amplifying transistor or tube.

We can define noise figure as a power ratio in decibels. The equation for noise figure is

$$\mathrm{NF} = \frac{S/kTB}{S_o/N_o}$$

where NF is the receiver noise figure, S equals the signal input power, S_o is the signal output power from the receiver, and N_o is the receiver noise power. The term k is Botlzmann's constant (1.38×10^{-23} joules per degree Kelvin). T equals the absolute temperature of the signal source, and B is the receiver bandwidth. kTB is the noise power available from a resistor of specified value at a temperature, T. We can learn from this that any signal will have a minimum amount of kTB noise associated with it.

If it were possible to route a signal through a receiver without adding noise to it, the NF would be unity. Therefore, because a receiver adds noise to the signal by some amount, the NF will be greater than unity. A communications receiver of good quality, designed for reception up to 30 MHz, will usually exhibit a noise figure of 5 to 10 dB. Owing to the typical levels of man-made and atmospheric noise in the HF spectrum and lower, this noise figure is entirely acceptable. However, at frequencies in the VHF range and higher, the fundamental noise source is the amplifying devices in the receiver. This requires greater attention to the receiver noise figure. A well-designed VHF, UHF, or microwave receiver will have a noise figure substantially lower than 2 dB in order to ensure optimum sensitivity.

Noise figure can be measured easily by employing the standard equation

$$\mathrm{NF} = 10 \log_{10} \frac{N_2}{N_1} \quad \mathrm{dB}$$

where N_1 is the noise power in watts from the output of an ideal receiver for a temperature of 290° Kelvin (16.82°C), and N_2 is the noise-power output from the receiver under test at the same temperature.

In a practical situation where we would measure the noise figure by means of a noise generator, the characteristic input impedance of the receiver must be interfaced with a like impedance at the noise-generator output. Typically, this will be 50 Ω to 50 Ω for modern receivers. The procedure is to measure the receiver noise output first, with the noise generator inoperative. An arbitrary noise reference level is chosen in decibels. The noise source is turned on and its output level is elevated until a convenient power ratio is expressed by N_2 over N_1. Once these numbers are obtained, we can calculate the noise figure by

$$\mathrm{NF} = -10 \log \left[\frac{N_2}{N_1}\right] - 1$$

If a diode type of noise generator is used, the excess noise created by the generator can be determined by

$$N\ (\text{excess}) = 10 \log (20R_d I_d)$$

where N is the excess noise in decibels, R_d is the output resistance of the noise generator, and I_d is the diode current in amperes.

The importance of low-noise amplifiers is particularly significant in the early or low-level stages of a receiver or composite amplifier. It is standard practice to employ low-noise transistors of ICs in the first stage of IF and even audio amplifiers to ensure a satisfactory noise figure.

Although for maximum stage gain in an RF amplifier it is imperative to strive for an impedance match between the signal source and the amplifier, and between the amplifier and its load, this does not necessarily yield the best noise figure. It is common practice, especially at VHF and higher, to trade gain for optimum noise figure by introducing an intentional mismatch. Also, the operating parameters of the amplifiers (biasing) are often varied until the noise figure is the lowest. If a gain trade-off is anticipated, it is prudent to select a transistor that has an f_T that is significantly higher than the operating frequency. This is because a bipolar transistor provides an increase in gain of approximately 6 dB each time the frequency is moved one octave lower. Therefore, a surplus of initial gain is available at the chosen operating frequency as referenced to the manufacturer's rated gain at some specific operating frequency for a particular transistor. We must realize that a consequence of this practice may be that of amplifier instability. Self-oscillation or regeneration of this type is a frequent cause of poor noise figure in an amplifier, so adequate measures must be taken to assure that such instability does not occur.

3.3 AMPLIFIER STABILITY

Not only can an unstable amplifier cause a degradation in noise figure, it can produce a host of spurious frequencies that can mix with the desired signal and produce additional spurious responses above and below the desired frequency. Responses of this variety can cause "birdies" in a receiver, or spurious emissions from a transmitter. In a worst-case situaation, the self-oscillations can be strong enough to block or impair the dynamic range of a succeeding stage in the circuit. If the peak-to-peak voltages developed during the oscillation period are great enough, destruction of the transistor can occur. This phenomenon is especially critical in RF power amplifiers that contain transistors with maximum V_{ceo} (collector-to-emitter voltage, base open) ratings that are not at least three or four times greater than the V_{cc} (supply voltage). As a minimum design requirement, the transistor chosen for a circuit should have a V_{ceo} rating of at least twice the V_{cc} for CW applications. This will allow the necessary margin for the sine-wave excursion when signal energy is present.

Self-oscillation can take place anywhere in the frequency spectrum, depending upon the circuit conditions that encourage instability. Oscillations often become manifest from audio frequencies through the VHF region. Oscillations at audio and low frequency are usually encouraged by the 6-dB/octave gain increase discussed in Section 3.2.1. Layout problems, RF ground loops, and inadequate decoupling of the stages are the usual causes of instability near or above the operating frequency. An ideal amplifier would be unconditionally stable, it would not self-oscillate or be regenerative with or without feedback networks or input-output loads. The ideal performance of such an amplifier can be closely approximated if care is exercised during the design and construction of the circuit.

3.3.1 Stabilization Techniques

Stability enhancement at or near the amplifier operating frequency can often be realized by using neutralization. This procedure calls for balancing out the unwanted feedback by including another feedback path that contains equal-amplitude voltage of the opposite phase. This neutralizing voltage is taken from the amplifier output and fed back to the input port. We can see the mechanics of this technique in Fig. 3.10. C_f represents the internal positive feedback component of the amplifier. C_n provides neutralizing voltage that is 180° out of phase

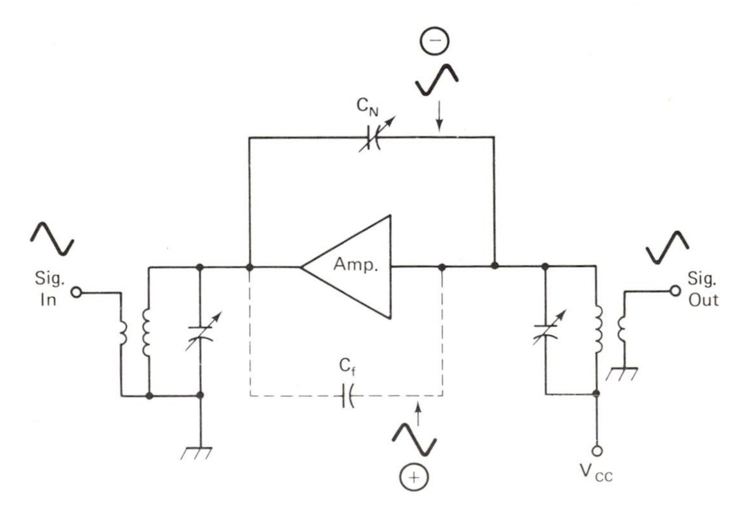

Figure 3.10 Representation of feedback and neutralization voltages in an RF amplifier.

with C_f, thereby canceling the internal feedback when the amplitudes of the two voltages are equal, resulting in a sum of zero. When this condition is reached, the amplifier responds as if both feedback paths were removed. Neutralization is effective only when the unwanted feedback path is within the amplifier. Stray feedback around the amplifier via unwanted circuit coupling must be treated by careful layout techniques and RF shielding that isolates the amplifier input from the output.

Neutralization over a broad range of frequencies is extremely difficult with the transistors as compared to vacuum tubes. This is because the input and output capacitances of transistors vary with frequency and junction current. Therefore, neutralization is likely to be most effective when the amplifier is designed for a narrow band of frequencies and when the operating current and voltages are relatively constant.

A more elegant form of neutralization is called *unilateralization*. This technique dictates the inclusion of a resistance in series or shunt with the neutralizing capacitor. The resistance helps to match the phase shift in the positive feedback path, caused by the resistive components within the transistor. The value of resistor used is dependent upon the operating frequency, with 100,000 Ω being somewhat typical at medium frequency, such as the standard AM broadcast band. As the operating frequency is increased, the resistor will have a smaller ohmic value. Generally, the best value for the resistor will have to be determined empirically. In any event, once the amplifier is neutralized or has been subjected to unilateralization, the tolerances and adjustments will be valid only for the transistor used during the stability adjustments; a replacement transistor of the same brand and type number can have internal capacitances that differ from those of the original device by as much as ±20%.

A brute-force approach to stability can be carried out by simply reducing the stage gain. With this method the unwanted positive feedback will still exist, but the gain will be lowered to a level where the feedback will not cause self-oscillation. Unfortunately, the gain of the stage will be substantially lower than that of a similar neutralized amplifier, often to a point of being unacceptable to the designer.

Although the circuit in Fig. 3.11 represents an overkill in stability measures, it includes for the purpose of our discussion the most common preventive measures. VHF and UHF parasitic oscillations can be killed effectively by insertion of parasitic suppressors at the input or output ports of the transistor. A small ferrite bead, labeled F.B. can be installed at the base or collector terminal near the transistor. The bead should have a μ_i (initial permeability) of 125 for frequencies of amplifier operation between 10 and 30 MHz. A 950 μ_i bead can be used below 10 MHz.

A cure for VHF or UHF self-oscillations can be effected also by the insertion of a small noninductive resistance at the base or collector

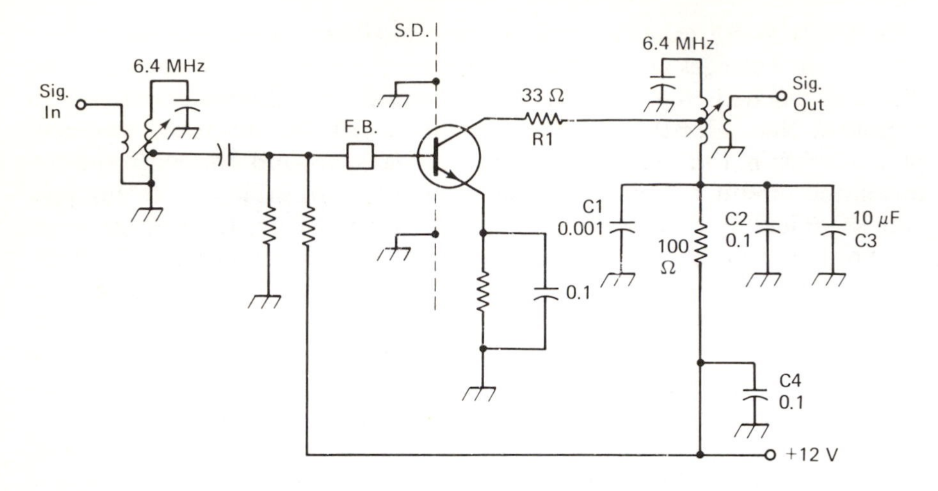

Figure 3.11 Practical example of a bipolar-transistor RF amplifier with tapped-coil impedance matching.

of the transistor. This is seen as $R1$ of Fig. 3.11, where a 33-Ω resistor is indicated. Generally, resistance values from 10 to 100 Ω are used for this purpose. A good rule is to use the lowest amount of resistance that will provide the desired damping action. Ferrite beads and parasitic resistors are equally suitable when used with FETs and ICs.

In especially stubborn cases of self-oscillation we may find it necessary to include a shield divider (labeled S.D. in Fig. 3.11) to isolate the input of the amplifier from the output portion. This will be helpful when the feedback is occurring outside the transistor rather than internally. Stray coupling between the tuned circuits and related components is the usual cause of this form of instability, but RF currents from the output of the amplifier can also reach the input circuit via the circuit board. It is an established fact that RF currents do not like to cross 90° barriers, so a piece of brass or copper shielding can be effective across the bottom of the transistor, as shown, when installed at right angles to the plane of the circuit board or chassis.

Decoupling of the amplifier V_{cc} line is also vital to stable performance. $R2$, in combination with $C1$, $C2$, $C3$, and $C4$ of Fig. 3.11, comprise an effective RC network for our purpose. Various capacitor values are specified. This is to provide effective bypassing at VHF, HF, and the lower frequencies. A 100-Ω resistor is used at $R2$ and is a standard value for most small-signal amplifier decoupling circuits.

Sometimes a designer will introduce an intentional mismatch in the interest of stability. This can be done by choosing the tap points on the input and output tuned circuits in an empirical fashion. The penalty for this rather severe stability measure is a loss of amplifier gain, as we learned earlier in this section.

3.4 PRACTICAL CIRCUITS

This section will cover a variety of practical, proven small-signal amplifiers that employ bipolar transistors and FETs. Some approximations are included for the purpose of arriving at a suitable starting point for a design. Final optimization can be achieved during the laboratory test period by varying the operating parameters. This form of finalization is by no means uncommon in the industry, despite careful initial design by means of mathematical procedures. This is because the exact characteristics of a given transistor are seldom known when it is first connected to a circuit. Variations of considerable magnitude are found in a given production run, which makes it virtually impossible to achieve an optimum design on paper. Our concern is mainly one of β and f_T for bipolar transistors and for the forward transconductance of field-effect transistors. There can also be a significant spread of internal capacitances among the transistors from a single production run.

3.4.1 Neutralized FET RF Amplifier

A JFET 100-MHz amplifier is presented in Fig. 3.12. $C1$ is used for input matching and noise-figure optimization. Its setting, plus the tap point on $L1$, are variables that can be adjusted for the desired operating condition.

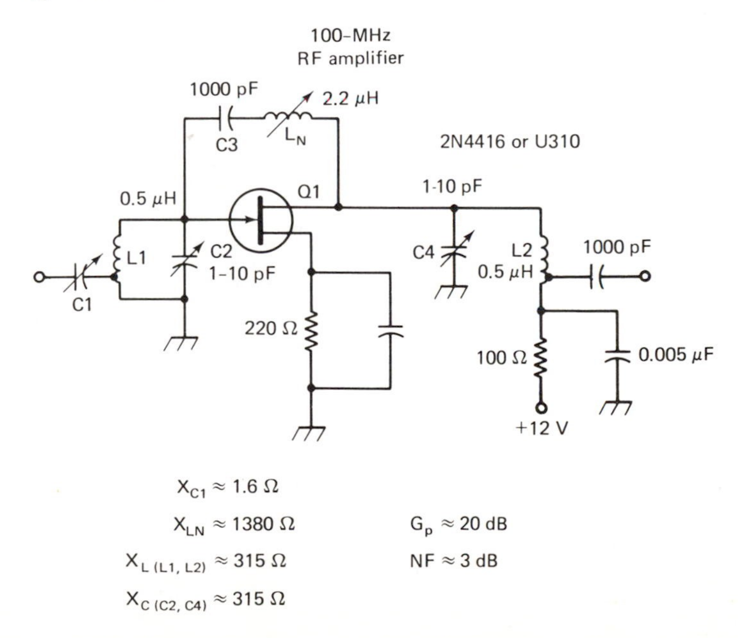

Figure 3.12 Neutralization can be effected by using a variable inductive reactance between the drain and base of an FET amplifier.

$C2$ is usually a blocking capacitor with low X_c, as in the case of a bypass capacitor. L_N is a neutralizing reactance with an adjustment slug for precise setting during the neutralizing process. Its setting will also have a noticeable effect on the amplifier noise figure. Resonator $C4/L2$ should be adjusted before neutralization is undertaken. There is usually some interaction between the drain tuned circuit and L_N, requiring re-adjustment of both elements during amplifier alignment. Input and output coupling for this circuit can be achieved by means of small links rather than with the capacitive technique illustrated. If the latter is employed at $L2$, it can be tapped at the point of $L2$ that provides a proper output-impedance match to the load. $Q1$ can be a Siliconix U310 in the interest of high dynamic range and low noise figures. A Motorola MPF102 or 2N4416 is satisfactory for most nonstringent applications. a 500-Ω potentiometer can be substituted for the fixed-value source-bias resistor during the initial optimization period. It will permit the designer to find the best bias point for noise figure.

3.4.2 JFET Cascode RF Amplifier

Good stability and high input and output impedance are the basic traits of the cascode amplifier. A suitable circuit for VHF use is provided in Fig. 3.13. A Siliconix U421 dual JFET is used at $Q1A$ and $Q1B$.

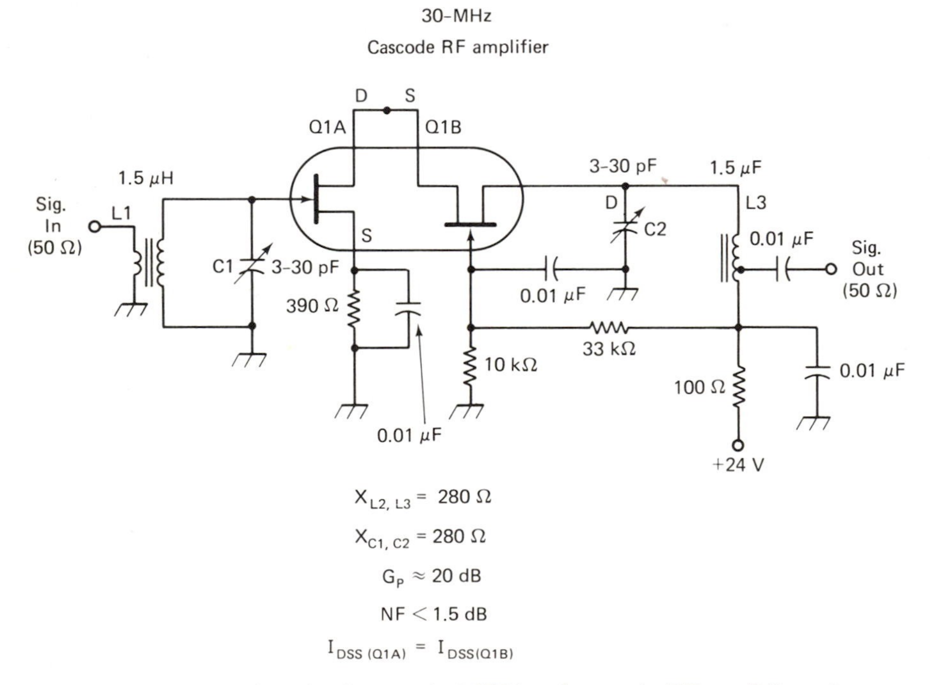

Figure 3.13 Circuit of a practical HF-band cascode RF amplifier using a dual-FET active device.

Good dynamic balance is assured by virtue of both FETs being formed at the same time on a common substrate. The conditions for initial design are listed below the circuit. If careful layout is used, the cascode amplifier will perform in a stable manner.

3.4.3 Cascode Amplifier with AGC

Discrete JFETs are shown in a cascode VHF amplifier that has AGC capability. A pair of 2N4416s are indicated at *Q*1 and *Q*2 of Fig. 3.14, although other FETs with equivalent characteristics could be

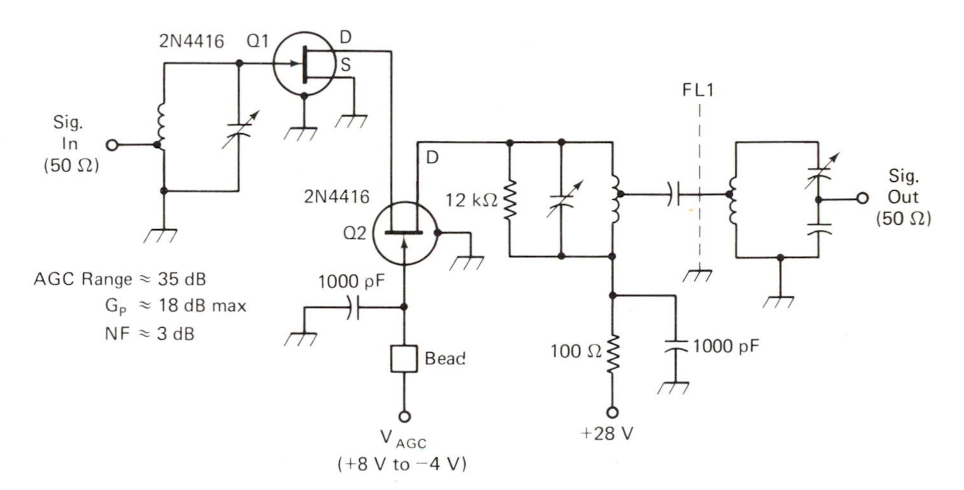

Figure 3.14 VHF cascode amplifier using discrete transistors and a bandpass filter.

used. An AGC control range of approximately 35 dB is typical with this amplifier. The noise figure is reasonably low, and the gain is on the order of 18 dB.

AGC control voltage of - 4 to +8 is applied to the gate of *Q*2. A ferrite bead of 125 μ_i is located directly at the gate terminal of *Q*2. The bead and the related 1000-pF bypass capacitor place the gate at RF ground and decouple the AGC bus to prevent interstage feedback, which can cause instability. *FL*1 is a bandpass filter that can be used to prevent out-of-band signals from passing to the succeeding stage, such as a mixer. If this is not a significant consideration, a resonator such as that shown in Fig. 3.13 will suffice.

3.4.4 Dual-Gate MOSFET with AGC

A different approach to applying GC to a dual-gate MOSFET amplifier is shown in Fig. 3.15. *D*1 is an LED that serves as 1.5-V refer-

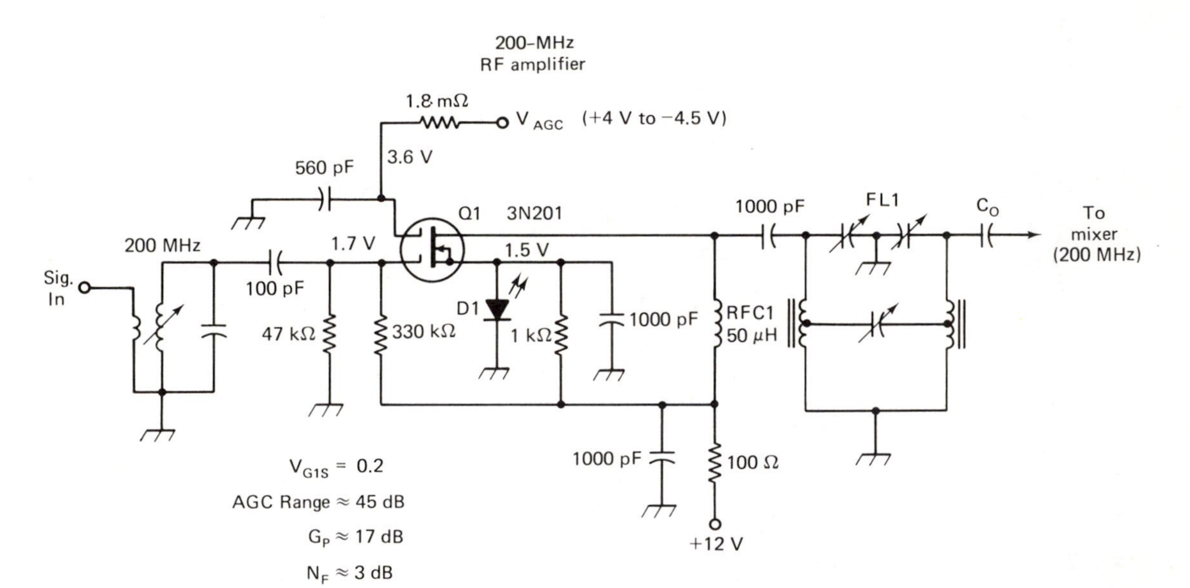

Figure 3.15 Practical dual-gate MOSFET RF amplifier for VHF, arranged for 45 dB of AGC control.

ence. Forward bias is applied to gate 1 from the V_{DD} line, and approximately 45 dB of AGC control is effected by applying -4.5 to +4 V at gate 2. At 200 MHz the noise figure will be near 3 dB and the gain will be roughly 17 dB, inclusive of the losses through $FL1$. C_o represents the load capacitance of the following stage and must be absorbed as part of the bandpass filter, $FL1$.

3.4.5 Broadband RF Amplifier

Some communications receivers are necessarily designed for high dynamic range. This is a notable objective we might pursue when designing a receiver for use aboard a military vessel, where several transmitters might be operating on a host of frequencies at a given time. The presence of large amounts of RF energy at the receiver input could, if an ordinary front end were used, lead to severe IMD (intermodulation distortion) and desensing of the early stages of the receiver. A solution can be had by employing a "strong" RF amplifier that is capable of high dynamic range in the presence of strong signals.

A practical approach to an RF amplifier that is capable of high dynamic range (linearity) is shown in Fig. 3.16(a). The principle of its operation is founded on the use of a class A bipolar transistor in which a substantial amount of quiescent current is permitted to flow. A 2N5109 is specified for $Q1$. This is a special low-noise device that was designed for this type of application in CATV amplifiers. It has a high f_T (1200 MHz), 2.5-W dissipation, a typical NF of 3 dB at 200 MHz, and a voltage gain of 11 dB at 150 MHz in a broadband configuration. The cross-modulation characteristic is typically -57 dB for +54 dBmV[b].

The circuit of Fig. 3.16(a) is arranged for degenerative and shunt feedback. The standing collector current is set at 50 mA. Input and output impedances of the amplifier are 50 Ω with the component values given. Selectivity is obtained through the use of bandpass filters at the input and output ports of the amplifier. $T1$ effects an impedance transformation from a 200-Ω collector characteristic to the 50-Ω filter, $FL2$. Another feature of this amplifier is excellent stability, even when no load is connected to it.

The general details of a suitable filter for use at $FL1$ and $FL2$ are provided in Fig. 3.16(b). Toroid cores (Amidon Associates T68-6 or equivalent) are used for the inductors. The primary of $T1$ and the secondary of $T2$ are wound to provide a match between the filter, its load, and its source.

3.4.6 VMOS FET Broadband RF Amplifier

A linear, high-dynamic-range RF amplifier was described by E. Oxner in *QST Magazine* for April 1979. It used a VMP-4 power FET in

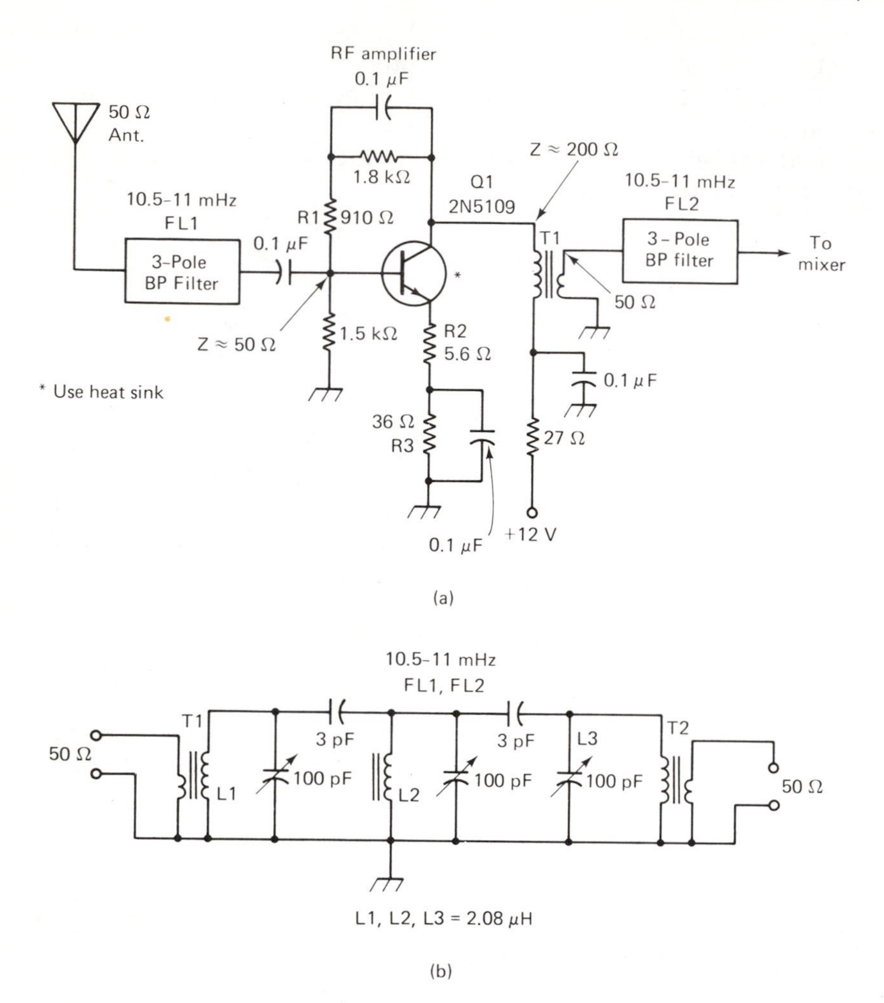

Figure 3.16 Details of a high-dynamic-range RF amplifier with filtering. A CATV 2N5109 is used to provide excellent linearity.

a broadband circuit. The virtues of VMOS devices are no thermal runaway, immunity to damage for mismatch, low noise, high g_m, and constant input and output capacitances, irrespective of applied signal and operating frequency. These characteristics make the VMOS FET ideal as a receiver RF amplifier when high signal-handling capability is desirable.

A version of the Oxner circuit is shown in Fig. 3.17, but it uses a less-expensive VMOS transistor, a Siliconix VN66AK. Power gain is approximately 13 dB, and the noise figure is 4 dB at 30 MHz. *RC* coupling is used at the amplifier input, but a bandpass filter (see Fig. 3.16) could be used if desired. Filtering can be included at the amplifier output also. As shown, the circuit contains a toroidal broadband

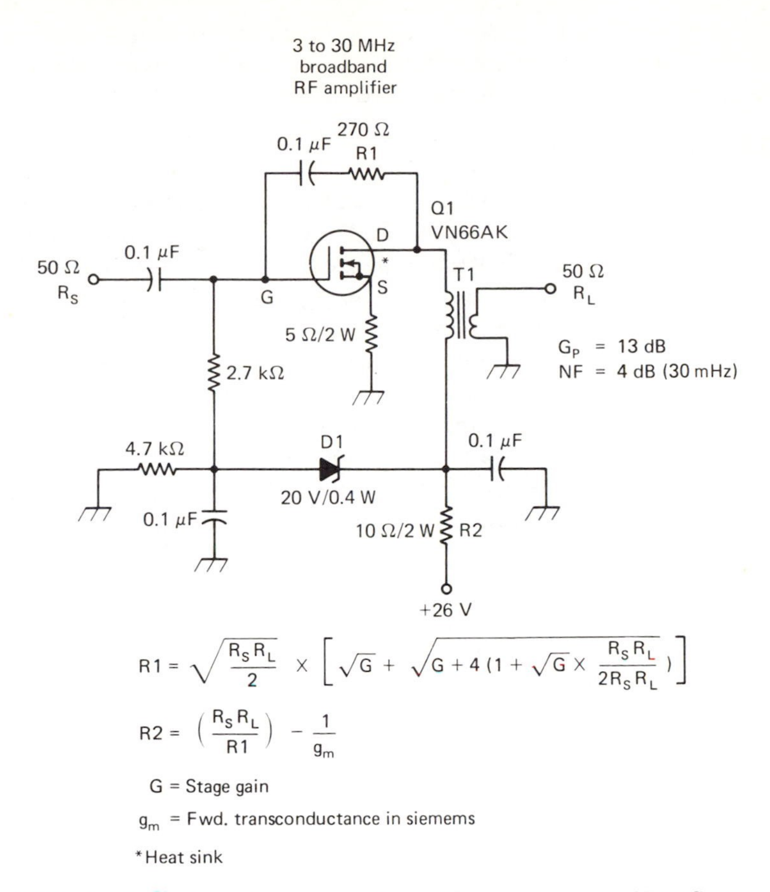

Figure 3.17 High-performance RF amplifier containing a VMOS power FET. The circuit is an adaptation of one designed by E. Oxner of Siliconix, Inc.

output transformer, $T1$. The primary of $T1$ consists of 12 turns of number 28 enameled wire on an Amidon FT-50-61 ferrite core ($\mu_i = 125$).

Correct values for $R1$ and $R2$ can be determined by means of the equations contained in Fig. 3.17. The known factors must be the g_m in siemens and the desired gain of the amplifier in decibels. Because a fairly large amount of drain current flows in the circuit, $Q1$ will require a heat sink. As an aid to linearity, the bias applied to the gate of the VN66AK is established through series regulator $D1$. Degenerative feedback is developed by means of the 5-Ω source resistor (unbypassed).

The secondary winding of $T1$ is wound to provide a suitable impedance match to the load that follows the amplifier. The drain impedance of this circuit can be approximated by

$$Z_D \approx \frac{V_{DS}}{1.3 \times I_D} \quad \Omega$$

where Z_D is the drain impedance, V_{DS} is the drain-source voltage, and I_D is the drain current in amperes. Thus, if the V_{DS} was 24 V and the I_D was 50 mA, the Z_D would be approximately 370 Ω.

3.4.7 Video Amplifiers

We should include video amplifiers in this discussion, since they are indeed RF amplifiers. In fact, the circuit described in Section 3.4.6 is well suited to that application. The field-effect transistor is excellent for use in video-amplifier circuits because gain-bandwidth products beyond 250 MHz are easy to realize when using simple one- or two-transistor circuits. Not only is this important to the design of video amplifiers, but we must be concerned also about input and output capacitances of the amplifier, as they can limit the frequency response of the circuit and lower the input impedance. For the foregoing considerations it is advantageous to use a cascode circuit in a video amplifier, basically because it has a very low input capacitance. The input capacitance can be obtained from

$$C_{\text{in}} = C_{gs} + (1 - A_V)\, C_{dg}$$

where A_V represents the voltage gain from the $Q1$ gate to the $Q1$ drain (Fig. 3.18), which is unity. C_{gs} is the gate-source capacitance and C_{dg} is the drain-gate capacitance. C_{in} can be expressed also as $C_{\text{in}} = C_{iss} + C_{gd}$, where C_{iss} is the small-signal input capacitance short-circuit. Thus, if a given transistor has a rated C_{iss} of 6 pF and a C_{dg} of 1.5 pF, the C_{in}

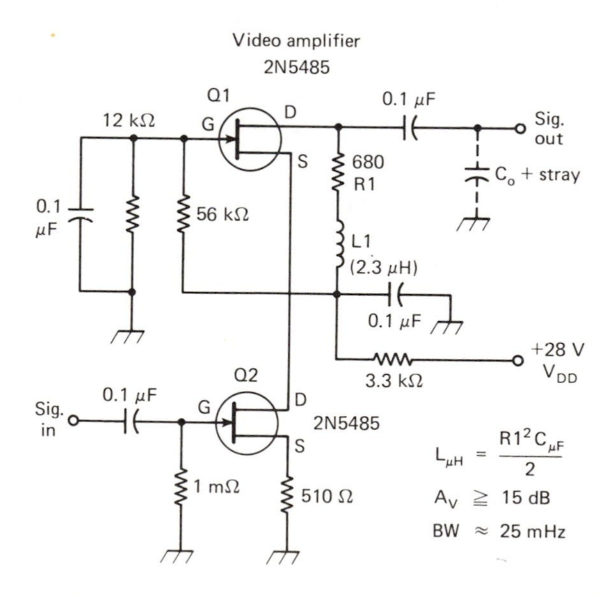

Figure 3.18 Video amplifier that uses FETs in a cascode arrangement for low input capacitance and good bandwidth.

would be 7.5 pF. We must take into account also the matter of stray circuit capacitance, which becomes part of the overall shunting capacitance. Typically, with good layout procedures, it is on the order of 5 pF. This would lead to an effective input capacitance (in the preceding example) of 12.5 pF. It is worthy of note that the input resistance of an FET is inversely proportional to the frequency squared. Conversely, the input capacitance of an FET does not vary up to at least 100 MHz. The same is not true of bipolar transistors, which makes a strong argument in favor of FETs for video amplifiers.

The output capacitance of a video amplifier becomes an important consideration when a shunt peaking coil is used, as in the circuit of Fig. 3.18. This is because the output C becomes part of the broadly resonant circuit formed by it and the peaking coil, $L1$. Although the peaking coil Q is very low, by virtue of the high series resistance ($R1$), it is desirable to maintain the output C as low as possible. In fact, the value of $L1$ is chosen in accordance with the actual and stray output capacitance by

$$L_{\mu H} = \frac{R1^2 \ C_{\mu F}}{2}$$

where $C_{\mu F}$ is the composite output capacitance, inclusive of stray C, as indicated by C_o in Fig. 3.18.

A practical example of a cascode video amplifier is provided in Fig. 3.18. Although discrete FETs are listed in the diagram, a Siliconix dual FET of the U257 type would be better in terms of dynamic balance. It also has a very low input capacitance.

A sometimes preferred video-amplifier arrangement is given in Fig. 3.19. A compound circuit utilizes an FET and a bipolar transistor in

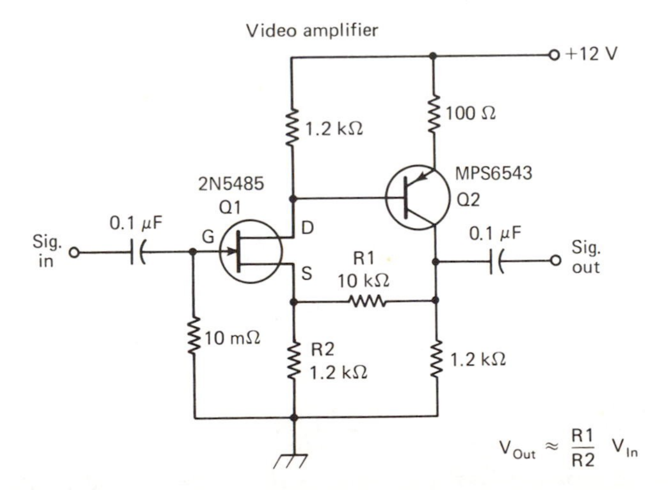

Figure 3.19 Compound video amplifier in which an FET and a bipolar transistor are used to provide high gain and low input capacitance.

cascade. This method offers excellent gain-bandwidth characteristics. A bandwidth of nearly 100 MHz was achieved with the circuit, plus a gain in excess of 10 dB was measured. If desired, a peaking coil can be inserted between the collector $Q2$ and the 1.2-kΩ collector resistor. The utility of this circuit is enhanced by the relationship of $R1$ to $R2$. Variable gain can be had by making $R2$ a potentiometer. A value of 2500 Ω is suitable. The simple equation in Fig. 3.19 expresses the gain relationship to the two resistances.

3.5 INPUT COUPLING TECHNIQUES

Earlier in this chapter we discussed the matter of impedance matching the amplifier to its signal source and load in order to ensure maximum gain. The exception was when an intentional mismatch was created to improve the RF amplifier noise figure or to aid stability through a reduction in gain.

When matching a bipolar transistor to a low-impedance signal source, it is a simple matter to tap the base on the tuned circuit at some point that represents the characteristic impedance of the base (generally 500 to 1500 Ω). But in the case of an FET amplifier this situation becomes more difficult because of the very high gate impedance of the field-effect transistor in a common-source circuit. Here we are dealing with impedances in excess of 1 MΩ, exclusive of the gate resistor we might elect to use.

A popular method for obtaining a particular impedance transformation is illustrated in Fig. 3.20. In the circuit of Fig. 3.20(a) we find $C1$ and $C2$ being used as a capacitive divider for the purpose of tapping the gate of $Q1$ down on the tuned circuit. This method is suitable for providing a matched condition, or it might be employed to cause an intentional mismatch, depending on the application. $L1$ serves in this example as the input coupling link over the low end of $L2$. Since $C3$ and $C4$ are used as bypass capacitors, their reactance should be less than 10 Ω at the lowest intended operating frequency. This will assure good bypassing action.

Absolute values for $C1$ and $C2$ are expressed more specifically in Fig. 3.20(b), where the capacitors are used to match R_{in} to R_L. It is necessary to first choose a bandwidth for the resonator. Once this is known, Q_L can be extracted from

$$Q_L = \frac{f}{f_2 - f_1}$$

where Q_L is the loaded Q of the resonator, f is the operating frequency, f_1 is lower frequency at the 3-dB point on the response curve, and f_2 is the upper frequency at the 3-db point on the response curve. Hence, if f

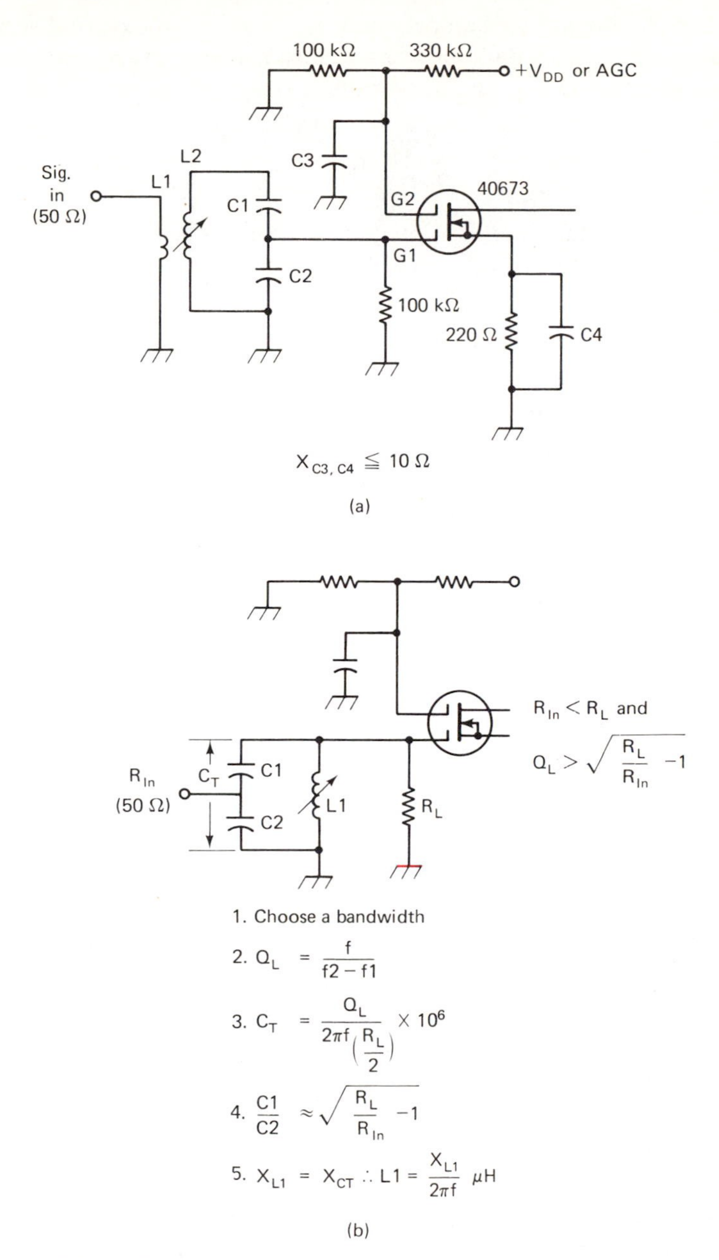

Figure 3.20 Capacitive dividers can be used effectively as impedance-matching networks at the input or output of amplifiers.

were 5 MHz, f_2 were 5.5 MHz, and f_1 were 4.5 MHz, the Q_L would be 5.

It becomes necessary to calculate next the C_T (total capacitance) required across $L1$ to effect resonance at 5 MHz. A value must be chosen for R_L before we can proceed. R_L can be an external resistor or the characteristic impedance of the transistor input terminal. Once this has been established, C_T can be obtained from

$$C_T = \frac{Q_L}{2\pi f \dfrac{R_L}{2}} \times 10^6$$

where C is in picofarads, f is in megahertz, and R is in ohms.

The single values of $C1$ and $C2$ are represented by

$$\frac{C1}{C2} \approx \sqrt{\frac{R_L}{R_{in}}} - 1$$

when R_{in} is less than R_L, and

$$Q_L > \sqrt{\frac{R_L}{R_{in}} - 1}$$

with R_{in} expressed in ohms.

Finally, since at resonance X_{L1} must equal X_{CT}, the inductance of $L1$ can be determined by

$$L1 = \frac{X_{L1}}{2\pi f} \quad \mu\text{H}$$

where X is in ohms and f is in megahertz.

BIBLIOGRAPHY

Application Notes

Hejhall, Roy, "RF Small-Signal Design Using Two-Port Parameters," Motorola AN-215A.

Maxwell, John, "The Low-Noise JFET—The Noise-Problem Solver," National Semiconductor Note 151.

Sherwin, James, "FET Biasing," Siliconix data file TA70-2.

Siliconix, "FET Cascode Circuits Reduce Feedback Capacitance," Applications Tip, August 1970.

Texas Instruments, "FET Design Ideas," Bulletin CB-145.

Texas Instruments, "TV Design Considerations Using High-Gain Dual-Gate MOSFETs," Bulletin CA-173.

Turner, Mike, "Why Use Cascode Dual FETs" National Semiconductor FET Brief 2.

Journals

DeMaw, Doug, "More Thoughts on Solid-State Receiver Design," *QST Magazine*, January 1971.

____ "His Eminence—The Receiver," *QST Magazine*, June and July, 1976.

Fisk, James, "Receiver Noise Figure, Sensitivity and Dynamic Range—What the Numbers Mean," *Ham Radio Magazine*, October 1975.

Hayward, Wes, "A Competition-Grade CW Receiver," *QST Magazine*, March and April, 1974.

Maxwell, J., "Hold Down Noise with JFETs," *Electronic Design*, February 16, 1976.

Reisert, Joe, "Ultra Low-Noise UHF Preamplifier," *Ham Radio Magazine*, March 1975.

Rohde, Ulrich, "Eight Ways to Better Radio Receiver Design," *Electronics*, February 20, 1975.

Sherwin, J. S., "Liberate Your FET Amplifier," *Electronic Design*, May 1970.

Technical Books

ARRL Radio Amateur's Handbook, receiving chapter, ARRL, Inc., Newington, Conn., 1980, 1981.

Cowles, J., *Transistor Circuit Design*, Prentice-Hall, Inc., Englewood Cliffs, N.J., 1972.

Hayward, W., and D. DeMaw, *Solid-State Design for the Radio Amateur*, Chapters 1 and 6, ARRL, Inc., Newington, Conn., 1977.

Rheinfelder, *Design of Low-Noise Transistor Input Circuits*, Hayden Book Co., Rochelle Park, N.J., 1964.

RCA, *Solid-State Servicing*, Chapter 2, RCA TSG-1673A, Somerville, N.J., 1973.

Sevin, L. T., *Field-Effect Transistors*, McGraw-Hill Book Co., 1965.

4

LARGE-SIGNAL AMPLIFIERS

Although we must be concerned with the parameters of voltage, current, beta, gain, stability, and impedance matching when working with RF power transistors, just as we are when using small-signal devices, other considerations must be added to the list. We now find ourselves confronted by the more significant complexities of input and output impedance with a somewhat different style of transistor geometry. Additionally, heat and the correct methods for controlling it must be pondered in the design process. The matter of transistor self-destruction needs to be addressed when designing for elevated power levels, lest an unreliable product result. We find also that nonresonant matching networks become more common than when dealing with low-level amplifiers. Our knowledge of causes and effects is perhaps of greatest importance when designing RF power applications for bipolar and VMOS transistors, and includes practical information on circuits, matching techniques, biasing, and stabilization. Spectral purity of the emissions is emphasized along with attention to linearity in amplifiers that require it.

4.1 NATURE OF RF POWER TRANSISTORS

Perhaps the worst problems we will encounter when working with RF power transistors are stability and damage to the device during operation. Manufacturers have taken important strides in semiconductor technology to minimize damage, such as the introduction of BET (balanced-emitter transistors) units by Motorola and "ballasted" transistors by RCA. In essence, these high-power transistors contain several bipolar devices on a common substrate. The collectors and bases are parallel-connected, respectively, at the time of fabrication, but the emitters each contain an internal small resistor (approximately 1 Ω). The opposite ends of the internal resistors are connected in parallel, then brought out to the emitter pin on the transistor header. By adopting this method, the designers were able to prevent "hot spotting" at any one transistor site on the substrate by virtue of current equalization provided by the individual resistors. Unfortunately, this type of transistor can still suffer catastrophic destruction under certain operating conditions, but short periods of mismatch, for example, can be tolerated, as is true during short periods of excessive voltage or current. The end effect of excessive voltage is punch-through of the junction, which causes an internal short-circuit. Excessive current (and heat) will generally cause the transistor junction to become open. Knowledge of these effects is helpful when determining the cause of a transistor burnout.

Instability can cause either of the foregoing problems. A self-oscillating transistor power amplifier can develop high peak voltages, often in excess of the safe ratings. Similarly, oscillations can increase the collector current to unsafe levels. Since stability is often frequency related, owing to the inherent 6-dB gain increase per octave lower, gain compensation becomes a vital factor when a transistor is operated well below its f_T, as in the case of broadband amplifiers. Furthermore, the phase relationships are conducive to instability at the lower frequencies (0.5 to 20 MHz) when the common-emitter mode is employed. Parasitic oscillations of this variety produce instantaneous energy that can destroy the transistors quickly. Low-level oscillations can, on the other hand, lead to gradual degradation of the semiconductor, with a progressive deterioration of the beta.

Damage can occur also at the operating frequency, irrespective of parasitic oscillations. This is typically the result of mismatch caused by removal of the load or a short-circuit in the load. A ballasted type of transistor can withstand short periods of mismatch, but a sustained mismatch will cause the transistor to exceed its P_D (maximum dissipation power) at a specified operating temperature.

Most RF power transistors have the inherent properties to make them act as varactors. This trait permits them to go into parametric oscillation at or near the operating frequency. For the most part, this form of oscillation can be prevented by means of careful physical construction and layout of the amplifier. Parametric oscillations are perhaps the worst kind in terms of unwanted spurious energy that can affect other communications services.

Since we are aware that gain and instability are closely related, we can recognize the pitfall of using a particular transistor at substantially less output power than it is rated for at a given frequency. This is because, as the collector current is lowered, the beta increases. For example, it would be imprudent to operate a transistor such as the Motorola MRF449A (rated at 13-dB gain for 30-MHz operation) at, say, 10 W of output. Not only is such a practice wasteful of the device capability and cost, but the stability could be a problem. The situation worsens markedly if the same device is operated at reduced power at some lower frequency, such as 2 MHz. Without excessive amounts of feedback to enhance stability, it would probably be impossible to tame the amplifier.

Most power transistors are rated at their *saturated power output*. This is the point where a further increase in driving power will produce no additional output power. Best stability generally occurs at that point on the curve, and the collector current should be at its rated amount, or nearly so, for the specified output power. In a practical situation, however, the transistor should be operated at 75% to 80% of the saturated output. This practice will lessen the "power slump" caused by temperature increases and will yield better amplifier efficiency. The greatest amount of saturated power output will be obtained from the common-emitter arrangement, because most of the driving power is fed through the amplifier to its output. There is no feedthrough of the driving power in the common-base configuration.

We must guard against *thermal runaway* in RF power transistors. This is a phenomenon during which the internal resistance of the transistor becomes less as the temperature increases. The reduced resistance gives rise to increased emitter-collector and emitter-base currents. Through this undesirable action, dissipation becomes greater and further lowers the internal resistance. These cumulative effects can destroy a transistor with rapidity. Adequate head sinking, which we will discuss later in the chapter, is required to prevent thermal runaway.

4.1.1 VMOS Power Transistors

Vertical metal-oxide silicon (VMOS) transistors, so named because of the V groove in their geometry, and referred to in some engineering circles as "V-FETs" or "vertical FETs," offer a number of advantages over bipolar devices in power amplifiers. Notably, they are not subject to *secondary breakdown* or thermal runaway. Secondary breakdown results from a situation of equilibrium during the thermal-runaway process, wherein there is a sharp drop in collector voltage. Intense localized heating usually destroys the transistor during this event. Furthermore, "current hogging" does not take place when more than one VMOS device is used in a parallel combination.

Power FETs are relatively immune to damage from mismatch while operating in the *on* mode. But a mismatch can lead to junction damage when the drive is switched while V_{DD} is applied to the amplifier or when the drive is constant and the V_{DD} is switched. These actions can create voltage transients under VSWR conditions, the peaks of which may exceed the safe ratings of the transistor.

Another desirable feature of the VMOS FET is its immunity to gain changes with frequency. As we learned earlier, this is not true of bipolar transistors. Additionally, the input and output capacitances (C_{in} and C_o) of power FETs are for the most part independent of the operating frequency. These two attributes make the power FET vastly superior to the power bipolar for use in fed-back, broadband amplifiers. The stability of the gain and capacitances make it a simple matter to design a feedback network that will function well over several octaves of frequency. Also, frequency-compensation networks are not required at the input of the power FET to level the amplifier gain. This is not true of wideband bipolar amplifiers.

It is important that we understand the fundamental nature of the two types of transistors under discussion. Figure 4.1(a) reveals that the bipolar device is *current driven*, whereas the VMOS FET of Fig. 4.1(b) is *voltage driven*. More simply, base current causes collector current, and gate-source voltage causes drain current. The FET gate is insulated from the source by a thin layer of silicon oxide. Ideally, there will be no current flow into the gate when V_{GS} is applied. However, leakage current on the order of nanoamperes will flow. The dc gain of the MOSFET device is meaningless, as the leakage current is all that can be considered, which translates into a current gain of roughly 10^9. Transconductance (g_{fs}) therefore becomes the meaningful parameter for the power FET. This is the change in drain current caused by 1-V change in gate-source voltage. Most VMOS power FETs designed for RF amplification have forward transconductances of 200 millisiemens (200,000 μS) or greater. This accounts for their ability to develop large amounts of output power with very low drive levels. This, then, is another advantage over the bipolar transistor. A typical VMOS FET will draw its rated drain current when a V_{GS} of 20 V peak-to-peak (p-p) is applied. An unfortunate by-product of the tremendous g_{fs} of this device is a tendency toward instability, especially at VHF. Therefore, we must be careful when laying out the amplifier, and we may find it necessary to employ one or more small ferrite beads at the transistor gate to discourage VHF parasitics.

Since essentially no current flows in the gate, the device input impedance is very high (as great at 1 MΩ). However, the exact impedance will be frequency related to some extent by the transistor input capacitance (C_{iss}), which for most medium-power VMOS FETs is on the order of 30 to 50 pF. Some high-power VMOS transistors exhibit substantially higher C_{iss} values, as well as high C_{oss} (output capacitance) amounts. The Siliconix VN64GA, for example, which is an 80-W device, has a

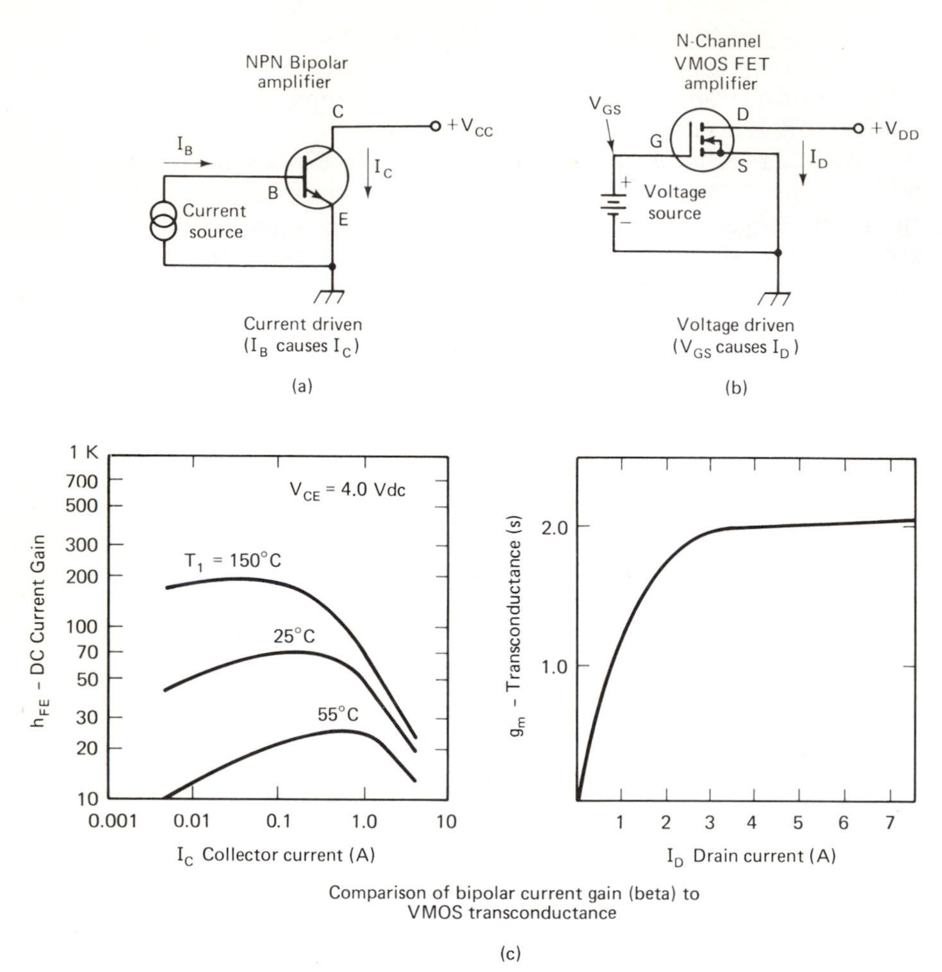

Figure 4.1 Functional comparison between bipolar and VMOS FET power amplifiers. The characteristic curves at (c) were provided courtesy of Siliconix, Inc.

C_{iss} of 700 pF and a C_{oss} of 325 pf. A VN66AK (8.33-W transistor), on the other hand, has a C_{iss} of 33 pF and a C_{oss} of 35 pF. When selecting a power FET, it is important to take these characteristics into account, since the capacitance becomes a consideration when designing the matching networks. Although these capacitances may seem unusually high compared to what we would encounter with power vacuum tubes, it is worth noting that it is not unusual to deal with bipolar-transistor input capacitances of 2000 pF or more in the HF spectrum.

VMOS devices offer greater efficiencies than can be realized with bipolar transistors. A typical class C bipolar-device amplifier will yield 50% to 60% top efficiency, whereas the writer has measured class C efficiences as great as 78% with VMOS FETs at comparable power levels. This level of efficiency, plus the general characteristics of the power FET, make VMOS transistors quite similar in performance to triode vacuum tubes.

It would be derelict not to mention the shortcomings of power FETs in this book. Notably, they are very intolerant to excessive V_{GS} and V_{DS}. Concerning the former, some power FETs contain built-in 15-V Zener diodes from gate to source. Although the diodes tend to protect the device from damage within specific overvoltage limits, they contribute to the effective C_{iss}. There has been a trend away from including the diodes during the manufacturing process, thereby increasing the upper frequency limits of the transistors. When protection is needed, it is a simple matter to install back-to-back Zener diodes from gate to source, externally. Figure 4.2 shows this technique. In the ex-

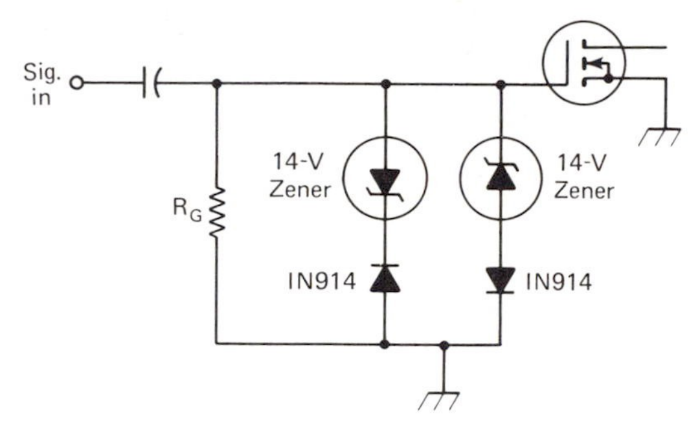

Figure 4.2 Zener diodes in combination with small-signal switching diodes provide gate protection for VMOS FETs.

ample, two 14-V Zener diodes are used in combination with two silicon high-speed switching diodes (1N914 or similar). By placing the two types of diodes in series, we can greatly reduce the resultant shunt capacitance. Some manufacturers specify a 30-V p-p maximum V_{GS}, but a 20-V swing is the practical limit when allowing a proper margin of safety. Within the 0- to 20-V range it should be possible to develop full power output from the FET.

Slight excursions beyond the maximum permissible V_{DS} ratings usually cause immediate destruction of the VMOS device. Such overvoltage conditions can result easily from self-oscillations, or spikes on the V_{DD} supply line. Zener diodes can be used effectively (back to back as in Fig. 4.2) between drain and source. They should clamp at some voltage slightly below the maximum acceptable V_{DS}. In the case of a 28-V V_{DD}, and allowing for the two times V_{DD} swing in an ac amplifier, the Zener diodes should be rated at approximately 58 V. This assumes a V_{DS} maximum of 60 V for the transistor used. Inclusion of Zener diodes in the gate or drain circuits is not recommended above 50 MHz because they exhibit unwanted shunt capacitance and may therefore complicate the impedance-matching network design.

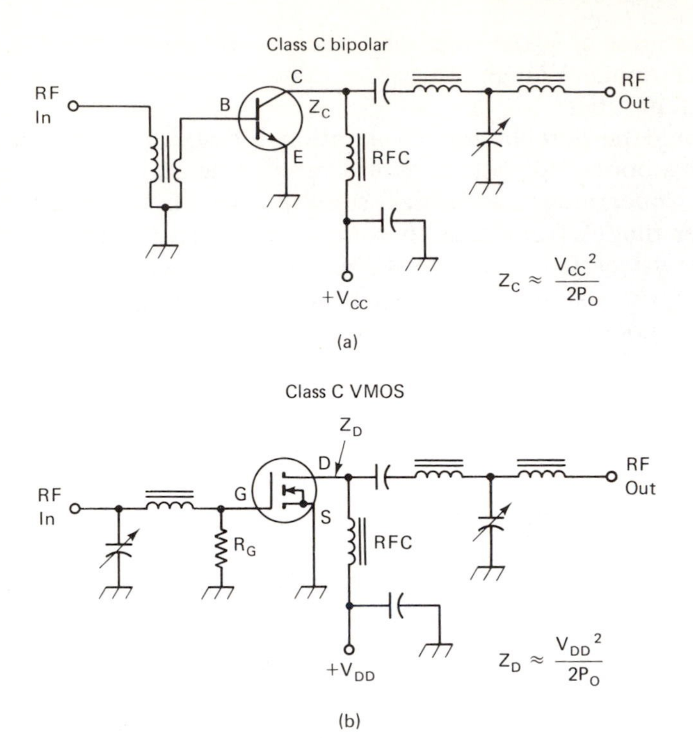

Figure 4.3 Class C amplifiers that use (a) a bipolar transistor and (b) a VMOS FET.

Figure 4.3 shows how a bipolar transistor and a VMOS power FET compare in terms of circuit arrangement for class C operation. Forward bias is not used in either example. R_G is selected for the circuit of Fig. 4.3(b) according to the driving power available. That is, there must be sufficient available drive to develop the required V_{GS} swing across R_G for full output power from the amplifier. In the interest of stability it is wise to use the lowest value of resistance possible at R_G. Values from 50 to 1000 Ω are typical.

The value of drain impedance is determined in the same manner as for a bipolar transistor collector impedance. A close approximation can be obtained from

$$Z_o = \frac{V_{CC}^2}{2P_o} \ \Omega \quad \text{or} \quad Z_o = \frac{V_{DD}^2}{2P_o} \ \Omega$$

where P_o is the anticipated output power in watts. A more precise determination requires knowledge of the actual collector-to-emitter voltage (V_{CE}) with drive applied to the amplifier. Similarly, we need to know the actual V_{DS} for a power FET in the interest of reasonable accuracy. This will be dependent on the $R_{DS(\text{on})}$ characteristic of the power FET, which is the resistance of the drain-source junction when the transistor is fully conducting. For a VMOS power FET this is on the order of 0.5 Ω, and we can use that number for most of our calculations. Now our equation for Z_o becomes

$$Z_o = \frac{[V_{DD} - V_{DS(\text{on})}]^2}{2P_o} \quad \Omega$$

To find the value of $V_{DS(\text{on})}$, we will consider it equal to

$$V_{DS(\text{on})} = R_{DS(\text{on})} \times I_D \text{ volts}$$

The results of these equations can be demonstrated by choosing some values for a hypothetical VMOS amplifier. Assume we apply a 24-V V_{DD} to a selected transistor, $Q1$. It will have an I_D of 3 A and a power output of 72 W. Thus for a R_{DS} of 0.5 Ω,

$$V_{DS(\text{on})} = 0.5 \times 3 = 1.5 \text{ V}$$

Therefore, the Z_o of the amplifier will be

$$Z_o = \frac{(24 - 1.5)^2}{144} = 3.5 \ \Omega$$

This will be the load line we use when designing our matching networks. VMOS amplifier efficiencies of 65% or greater are typical for the linear class A or B modes.

4.1.2 VMOS Class E Amplifier

Efficiencies as great as 90% can be obtained by operating bipolar or VMOS power transistors in the class E mode. The virtues of operating the devices in this switching mode are low power dissipation, high efficiency, insensitivity to component tolerances, and low junction temperature. Figure 4.4 shows a class E circuit that uses a Siliconix VMP4 power FET. The network Q_L for amplifier testing was 3. The equations in Fig. 4.4 are based on the papers by N. Sokal, referenced in the bibliography section of this chapter.

4.2 MATCHING NETWORKS

We acknowledged earlier in this chapter that maximum power transfer will occur only if the amplifier device is matched correctly to its source and load impedances. Therefore, it becomes an important design consideration to select networks that are suitable for this and the stability requirements of the amplifier. Fundamentally, the design approach re-

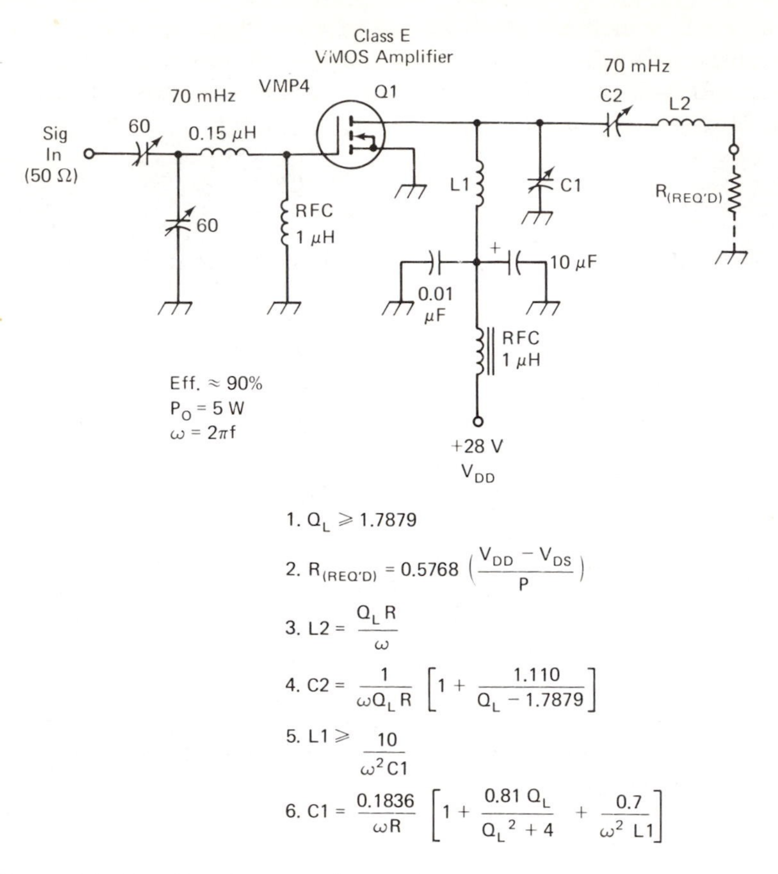

Figure 4.4 Details of a class E VMOS amplifier with network equations developed by N. Sokal of Design Automation Corp.

mains the same, irrespective of the class of operating mode. The notable exception is in the case of the class E amplifier described in Fig. 4.4.

Matching networks can be narrow band or broadband, depending on our requirements. The former usually consist of variable-*C* components and a fixed inductance, but in some instances the designer may elect to make both the *L* and *C* network components variable. This allows great latitude in effecting a matched condition when such unknown factors as input and output capacitance must be addressed. A totally adjustable network of this type is sometimes called a "sloppy network." Adjustment proceeds along purely empirical lines once the network values are arrived at by means of approximation. Unfortunately, the foregoing technique is the rule rather than the exception when working with power bipolar transistors, except for class E amplifiers, where a substantial amount of leeway is acceptable. Empirical adjustment is necessary because of the variations in transistor characteristics for various brands of the same transistor type number. Also, many data sheets do not provide meaningful information concerning the input and output capacitances, especially at various operating frequencies. This is not a problem with VMOS power FETs, and relates primarily to bipolar devices. The totally adjustable network allows the designer to compensate also for stray inductive and capacitive reactances that are always present in a solid-state amplifier circuit.

Most broadband amplifiers contain *LC* filters of the low-pass or bandpass variety. The wide-band filters are usually accompanied by broadband matching transformers of the toroidal kind. Other broadband amplifiers contain no selective filter elements and utilize nothing more than broadband transformers.

4.2.1 Narrow-Band Networks

There is an essential difference in the design approach between vacuum-tube and solid-state amplifiers when using narrow-band networks. In the interest of waveform purity in a tube type of amplifier, the general practice is to design the network for high loaded Q, typically between 10 and 15. A somewhat different situation prevails when working with semiconductors: owing to the very high gain of most modern transistors, lower network Qs become almost mandatory to ensure amplifier stability. Engineers have found that a maximum network Q of 5 is the most viable. The waveform is usually purified to an acceptable level by adding harmonic filters at the amplifier output. Often it is sufficient to use an inexpensive half-wave type of filter for this purpose.

The pi-network turns out to be impractical for use in most high-power solid-state amplifiers. Although network values can be reached by application of the standard equations, the resultant component values usually become impractical, especially as the operating frequency is increased to the high-frequency part of the spectrum and above. The inductance values often become a small fraction of a microhenry, and the capacitance values increase to proportions that make the use of variable capacitors beyond reason.

A more suitable approach is seen in the T-network of Fig. 4.5. This circuit provides the desired low-pass response and yields practical values of L and C, even at 30 MHz. The T-network is especially suited to low values of Z_C that must be transformed to higher values of impedance, such as 50 Ω.

The network equations are included in Fig. 4.5, along with the answers based on the circuit values noted in the drawing. In the example we have assumed an amplifier efficiency of 50% and a Q_L of 4. The operating frequency is 30 MHz, and I_C at peak power output is 2 A. C_o is the transistor output capacitance at 30 MHz. This characteristic is usually available from the manufacturer. $C1$ should have an X_c

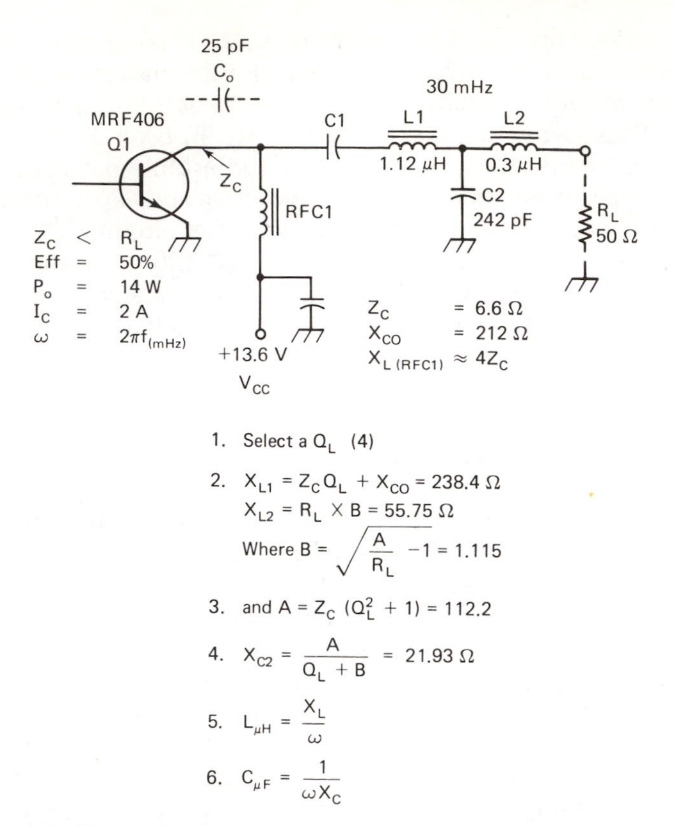

Figure 4.5 T-network design procedure for a collector matching circuit.

of 10 or less. Ceramic-chip capacitors are preferred at $C1$ in the interest of minimum series inductance and related phase shifts. $C2$ can be made variable to ensure precise adjustment of the network C in the presence of stray capacitance. This network is suitable for use at the input of $Q1$. $L1$ and $L2$ can be variable elements along with $C2$, hence permitting precise matching with the aid of an input VSWR meter.

A network that is particularly popular in VHF amplifiers, but works well at lower frequencies as well, is presented in Fig. 4.6. Like the circuit of Fig. 4.5, it is used when the Z_C is lower in magnitude than R_L. We can see that it closely resembles the conventional L network, except that $C2$ has been added to permit greater matching range and ease. A low-pass response is obtained from this network, an aid to the suppression of harmonic currents. Assume that the operating parameters for this amplifier are identical with those in Fig. 4.5. The reactances obtained from the equations of Fig. 4.6 are based on that premise. At the higher frequencies it is common practice to make both $C1$ and $C2$ variable. The values noted in the circuit for $C1$, $C2$, and $L1$ are the calculated values, but do not include stray C and L components.

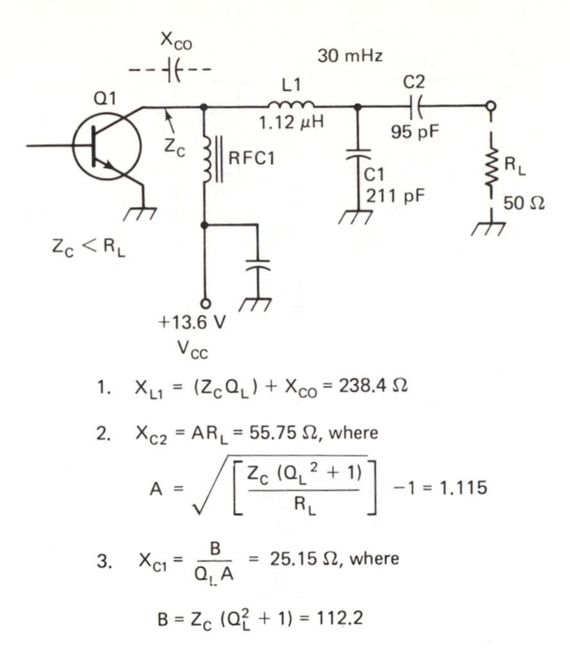

Figure 4.6 Design format for a modified L network used to match the amplifier to its load.

In situations where Z_C is greater in value than R_L, a pi network can be used effectively for matching purposes. This is shown in Fig. 4.7, where Z_C is 196 Ω and R_L is 50 Ω. Note that the operating conditions for this circuit are different from those of Figs. 4.5 and 4.6. V_{CC} has become +28 V, I_C is 0.143 A, and P_o is 2 W. This results in a collector impedance of approximately 196 Ω. In a practical circuit, $C1$ and $C2$ can be made variable.

The three networks described in this section can be used for interstage coupling and need not be used as shown. If the impedance levels are such that the equations provide impractical component values, a broadband transformer can be used between the transistor and the network to change the transformation ratio to a more workable value.

A tapped-coil narrow-band matching network can be an alternative to the three methods just described. This approach is seen in Fig. 4.8, where $L1$ is tapped for a low impedance to match the characteristic Z of the signal source. $C1$ and $L1$ form a resonator and provide the desired selectivity. $L2$ is wound over the ground or cold end of $L1$, using the proper number of turns to ensure a matched condition between the signal source and the base of $Q1$. The LC ratio provided by $C1$ and $L1$ is not critical as long as the parallel impedance of the tuned circuit is substantially higher than the two terminals being matched.

Although the same method is sometimes used at the amplifier output, greater matching range is offered through the use of two vari-

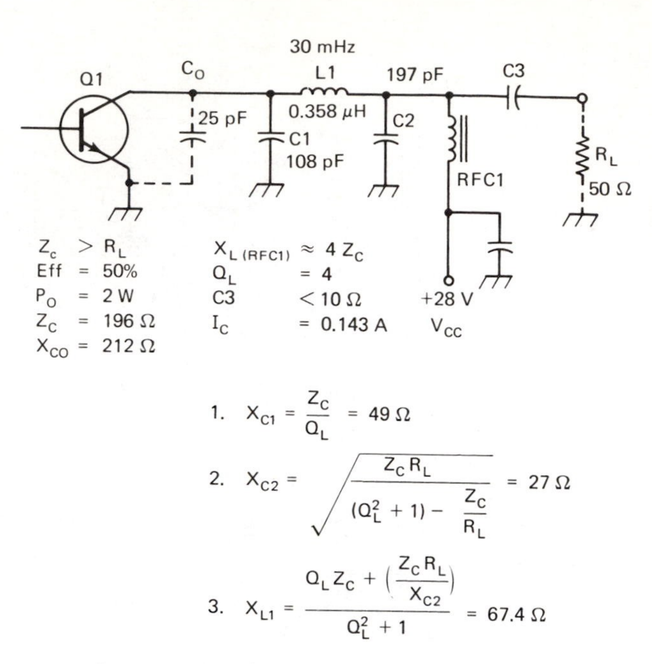

Figure 4.7 Pi-network design procedure for bipolar-transistor amplifiers.

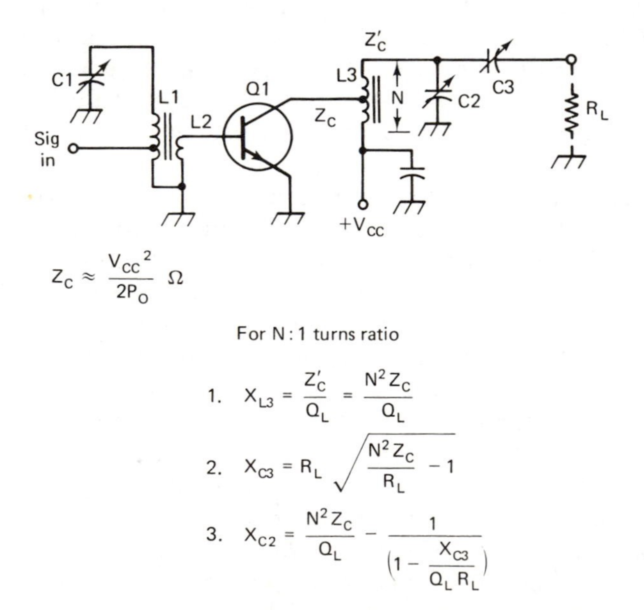

Figure 4.8 A tapped-coil matching network can be used with solid-state power amplifiers. Basic equations are provided here.

able capacitors ($C2$ and $C3$) and a coil tap. $C3$ of Fig. 4.8 will have the greatest effect on matching to R_L, but $C2$ will interlock with it sufficiently to require readjustment of both capacitors several times to effect a matched condition. The major limitation of the collector network, as shown, is poor harmonic suppression. This is particularly true when the tap point on $L3$ is near the top of the coil. The lower the tap location, the higher is the inductive reactance in the path of the harmonic currents, and hence the better the suppression. An advantage is, however, that with the collector of $Q1$ tapped down on $L3$, changes in transistor junction capacitance will have a minor effect on the tank-circuit resonance. Equations are included in Fig. 4.8 for determining the required values of X_L and X_C.

4.3 BROADBAND TRANSFORMERS

Most modern solid-state power amplifiers for use below 150 MHz employ broadband matching transformers. The suppression of harmonic currents is carried out by means of bandpass or low-pass *LC* filters at the amplifier output. Two kinds of transformers are used for impedance matching in broadband amplifiers. *Transmission-line transformers* are often used. They have windings that have a characteristic impedance and function electrically like sections of two-wire transmission line. The other type of transformer in vogue is the *conventional transformer*. There is some controversy about which type of transformer is best in terms of efficiency and high-frequency response, with the transmission-line transformer being favored in that respect. However, in practice we might find it difficult to differentiate between the two types without special measuring equipment. The losses in a conventional transformer, if they are of special concern, are so low that they are not important for most circuit work. Moreover, the conventional transformer offers the advantage of providing a match to almost any normal combination of impedances below approximately 600 Ω. The transmission-line transformer is limited to specific ratios, such as 4 : 1, 9 : 1, and so on.

Broadband transformers that are used below the VHF spectrum are wound on ferromagnetic cores, such as toroids or balun foundations. Ferrite core material is the usual choice, with the initial permeability (μ_i) running from as low as 40 to as much as 2500, with the higher μ_i being used at VLH. Powdered-iron core material can be used, but does not have the high permeability found in ferrite compounds; hence a given transformer wound on a powdered-iron core would require a much greater number of coil turns than a comparable transformer wound on a ferrite core of the same physical size. The trade-off is, however, a much lower flux density for ferrite, assuming a core of equivalent size to one made from powdered iron. Therefore, a ferrite core of a specified cross-sectional area will saturate more quickly than a core using powdered iron. Also, powdered iron is better for use in narrow-band tuned circuits above, say, 10 MHz. This is because the coil or tuned-transformer Q will be higher with the powdered-iron materials.

In principle, a broadband transformer covers a wide range of frequencies effectively because the core material plays a significant role at the lower end of the desired spectrum, but gradually "disappears" from the circuit (electrically) as the operating frequency is increased. Most HF-band broadband transformers are wound on ferrite cores that have a μ_i of 900 or 950, but permeabilities of 40 and 125 are commonly used for wide spectrums that start at approximately 10 MHz and extend upward in frequency. These can be considered useful rule-of-thumb guidelines.

In essence, we must consider the level of power our broadband transformer must handle, the μ_i, and the core losses. Also, the flux density (B_{op}) during operation must be within the linear portion of the BH curve.

When transmission-line transformers are used, we must be aware that the length of the transmission line needs to be great enough to provide the required number of turns on the magnetic core used. Conversely, it cannot be so long that it impairs the high-frequency response of the transformer. The line impedance for a transmission-line transformer can be determined by

$$Z_o = \sqrt{R_{\text{in}} R_L}$$

where R_{in} and R_L represent the terminal impedances of the transformer in ohms. In any event, the inductive reactance of the windings should be approximately four times the characteristic impedance of the terminals to which the transformer connects. Hence, if the transformer looks into a 10-Ω load, the inductance of the winding will be based on an X_L of 40 Ω. Therefore,

$$L_{\mu\text{H}} = \frac{X_L}{\omega}$$

where X_L is in ohms and $\omega = 2\pi f_{(\text{MHz})}$. Thus, if our lowest operating frequency was 2.3 MHz, and the transformer load was 10 Ω (40 Ω X_L), the required inductance to satisfy our winding requirement would be 2.89 μH. The routine procedure is to calculate the necessary inductance for the *smallest* winding of the transformer, as outlined previously. This ensures that the larger winding will have ample inductance to satisfy our rule.

4.3.1 Conventional Transformers

We can regard the conventional transformer as we might the audio or power transformer. That is, the windings are placed on the core separately, and the impedance ratio will be the square of the turns ratio. Therefore, if we needed a 16 : 1 impedance transformation, the transformer turns ratio would be 4 : 1.

Figure 4.9 contains a pictorial and schematic representation of a toroidal transformer of the conventional kind. Figure 4.9(a) shows a

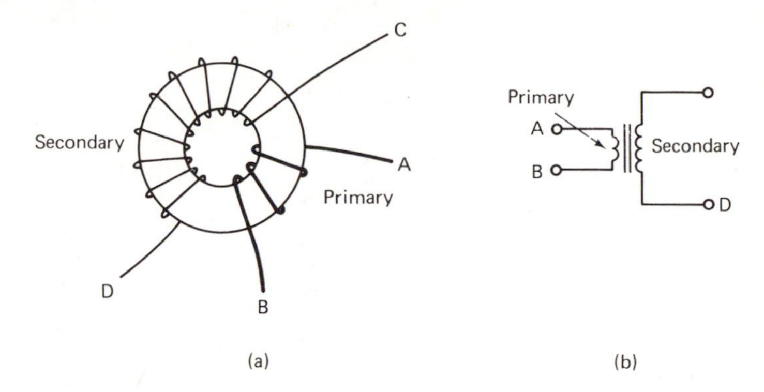

Figure 4.9 Pictorial and schematic representations of a broadband conventional toroid transformer.

larger winding on the left (secondary) and a small primary winding on the right. The electrical arrangement for this transformer is shown in Fig. 4.9(b). A magnetic-core rod could be used instead of the toroid, with nearly identical performance. The toroid form has the advantage of being self-shielding in nature, which is not true of a similar transformer wound on a rod or flat bar. In a practical situation the large winding would occupy all the core, and the smaller winding would be spread over the larger one. It was drawn as shown to simplify the illustration. Toroidal broadband transformers are most often used in low-level stages of transmitters and in small-signal circuits of other types of equipment.

A more common form of the conventional transformer is seen in Fig. 4.10. It is found in the base and collector circuits of solid-state RF power amplifiers and has the advantage of providing better physical and electrical symmetry than the toroidal style of transformer. In balanced circuits such as push-pull amplifiers, symmetry is important in terms of equal drive to each of the amplifier transistors.

The pictorial views in Fig. 4.10 show two headers (X and Y) that are fashioned from circuit-board material. Header X has solid copper on the outer side, to which the brass tubes are soldered to create attachment point E, the center tap of the primary winding. Header Y is made the same way, except for a groove down the center that has been etched out of the copper. This isolation groove permits connection points A and B to be electrically isolated from one another. The transformer secondary winding is looped through the copper tubes and exits from

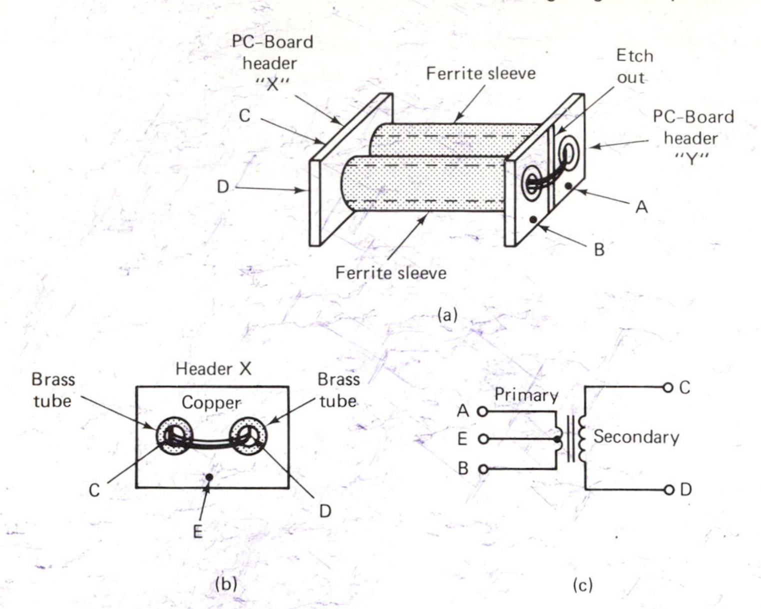

Figure 4.10 Structural details for a broadband conventional transformer that can be used as a combiner in a balanced circuit.

header X. The two lengths of brass or copper tubing constitute the primary winding, which is in effect a U-shaped half-turn. Since this transformer winding is so small, the ferrite material must have a high μ_i to provide the required winding inductance at the lowest operating frequency. A permeability of 950 is adequate for operation from 2 to 30 MHz in this style of transformer.

It is not imperative to use ferrite sleeves for the circuit of Fig. 4.10. Alternatively, two rows of high-mu toroid cores can be stacked and assembled as shown. The brass tubes and headers will hold them secure, assuming that the tubing fits snugly into the center holes of the toroids.

The transformation ratio is adjusted by adding or substracting turns from transformer winding C-D. Because of the physical character of this style of transformer, it is only possible to obtain even integers when developing a specified turns ratio. This type of broadband transformer lends itself nicely to direct mounting on a printed-circuit board by soldering the headers to the appropriate copper foils on the etched board. A good grade of insulation should be used on the secondary-winding wire, such as Formvar or vinyl. This will help prevent shorted turns on short-circuits to the headers.

4.3.2 Transmission-Line Transformers

Transmission-line broadband transformers contain windings that are formed by sections of two-wire transmission line. Often these windings consist of twisted lengths of magnet wire. The pair of wires can be chucked in a hand drill at one end and secured in the jaws of a vise at the opposite end. The drill is operated until the pair of wires has between six and ten twists per inch (25.4 mm). The characteristic impedance of the twisted pair will depend largely upon the wire gauge used. These bifilar windings need not be twisted, but can be laid on the transformer core as parallel conductors. The characteristic impedance will be approximately the same as for the twisted configuration. However, the twisted line is much easier to work with, and it will maintain its impedance more closely over its entire length than will a parallel-pair line, especially if small diameter wire is used.

If we were to construct a parallel-conductor bifilar winding of two lengths of number 30 (AWG) enameled wire, the resultant impedance of the line would be 32 Ω, as discussed by H. O. Granberg in *Electronic Design* for July 19, 1980. An identical pair of number 32 wires would provide a Z_o (characteristic impedance) of 62 Ω. This illustrates the relationship between conductor diameter and X_o. We could lower the impedance by twisting two or more pairs of wire together. To illustrate this let us assume we have wound four pairs of number 36 AWG together. The Z_o would now become approximately 18 Ω.

Most transmission-line transformers contain 25-Ω windings. For this purpose we can use subminiature 25-Ω coaxial cable such as Microdot 260-4118-000. Alternatively, two equal lengths of miniature 50-Ω line, such as RG-174/U, when placed in parallel, will yield a 25-Ω Z_o. The main limitation in using coaxial cable for the transformer windings is one of physical accommodation. Coaxial lines require larger core sizes than are needed for twisted or parallel sections of magnet wire.

Figure 4.11 shows how a 4 : 1 broadband transformer can be made from two baluns (balanced to unbalanced) to provide superb bandwidth (1.0 to 30.0+ MHz) and high-power capability. There are 14 turns of 25-Ω coaxial cable on each toroid core, but only 3½ turns are shown in order to simplify the illustration. External capacitors can be added to the input and output ports of the transformer to compensate for the leakage inductance. Trimmer capacitors are suggested for *C*1 and *C*2 of Fig. 4.12 to permit easy adjustment during the initial test period. Fixed-value capacitors can be substituted later to simplify production of large numbers of the transformers if they are to be used in a manufactured product.

If we analyze the circuit of Figs. 4.11 and 4.12, it will be apparent that the shield braid of the coax cable carries the high current of the low-impedance winding. This contributes to the overall efficiency of the transformer. For the frequency range this transformer accommodates we can use toroids of 950 μ_i, such as Micrometals type 43 of Indiana General Q1. Determination of the correct core size will be discussed later in the chapter.

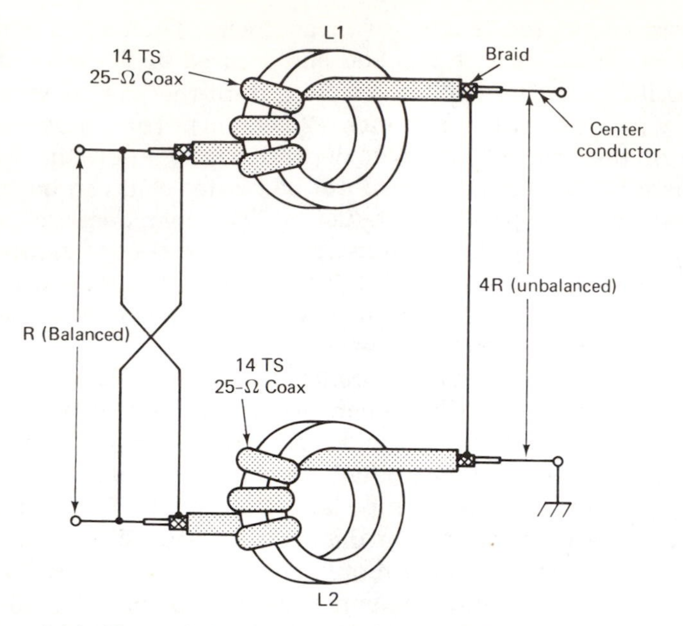

Figure 4.11 Transmission-line transformer technique suggested by Motorola Semiconductor Corp. for using miniature 25-Ω coaxial line sections. A 4 : 1 transformer is shown.

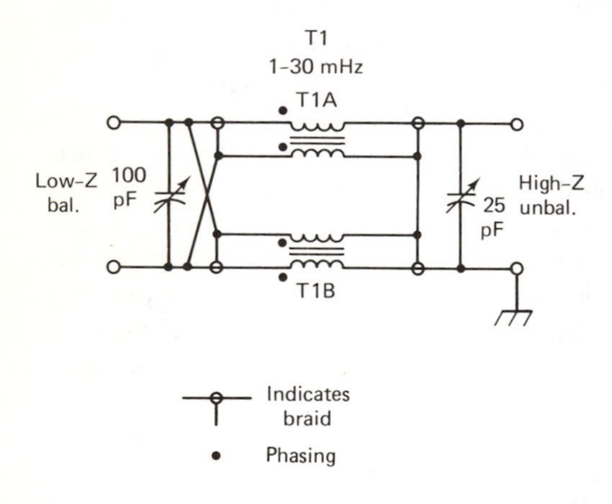

Figure 4.12 Electrical equivalent of the transformer shown pictorially in Fig. 4.11. Input and output trimmers are included for leakage inductance, an aid to high-frequency response. (M. F. "Doug" DeMaw, *Ferromagnetic-Core Design and Application Handbook*,©1981, p. 101. Reprinted by permission of Prentice-Hall, Inc., Englewood Cliffs, N.J.)

Four classic examples of transmission-line transformers are given in Fig. 4.13. Bifilar windings are used on the cores of all four transformers. Compensating capacitors may be required (as in Fig. 4.12) to extend the high-frequency response.

Additional configurations for broadband transmission-line transformers are suggested in Fig. 4.14. In Fig. 4.14(a) and (b) we find

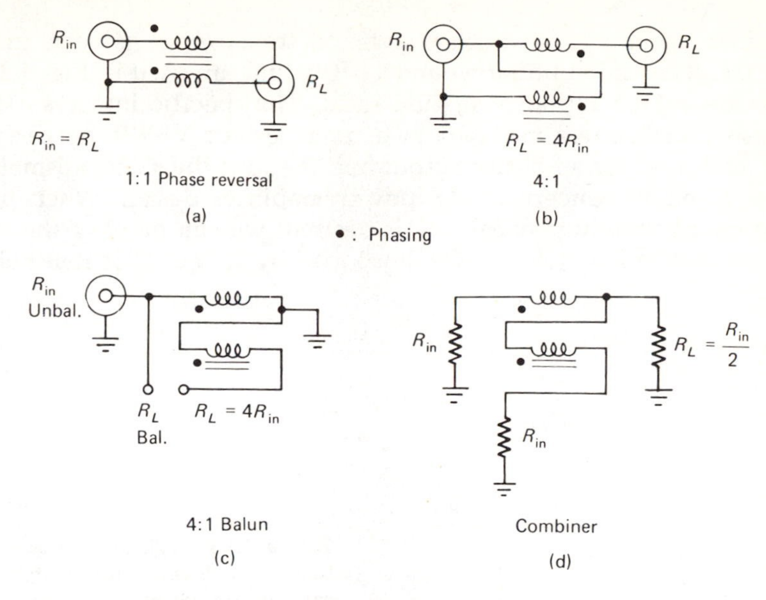

Figure 4.13 Circuit examples of various broadband transformers.

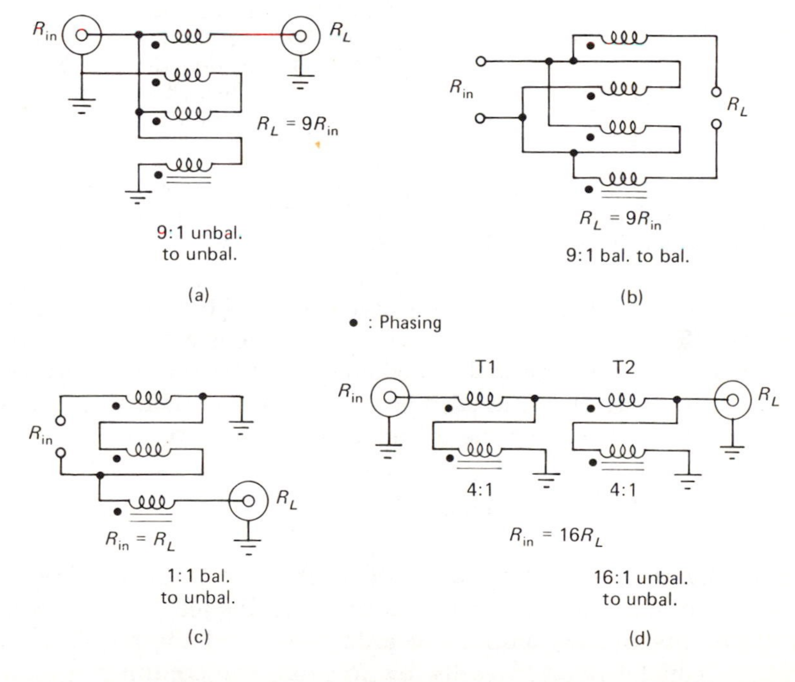

Figure 4.14 Transmission-line transformers that provide various impedance transformations. The windings are bifilar, trifilar, or quadrifilar, as indicated. (M. F. "Doug" DeMaw, *Ferromagnetic-Core Design and Application*, © 1981, p. 102. Reprinted by permission of Prentice-Hall, Inc., Englewood Cliffs, N.J.)

quadrifilar windings. A trifilar style of transformer is seen in Fig. 4.14(c), and cascaded bifilar-wound 4 : 1 baluns are used in Fig. 4.14(d) to provide a 16 : 1 transformation ratio. The specific integers offered by these transformers may not give us a perfect VSWR between the source and load, or amplifier output and R_L, but the slight mismatch is seldom a major concern in RF power amplifier design. When better resolution of matching capability is desired, we can employ the transformer shown in Fig. 4.15. It was developed by J. Sevick of Bell Laboratories. Additional X, Y, and Z taps can be added to the windings to secure ratios as low as 1.5 : 1. If a toroid core with a 2.5 in. (64-mm) diameter is used (125 μ_i for HF-band use) with number 14 or larger conductor size, the transformer should be capable of handling 1000 W of RF power safely, provided the VSWR on the line is kept below 1.5 : 1. This assumes that the terminal impedances seen by the transformer are quite low, say, 50 Ω or less at each port. A ten-turn quadrifiliar winding on a 125 μ_i core will be suitable for operation from 2 to 30 MHz.

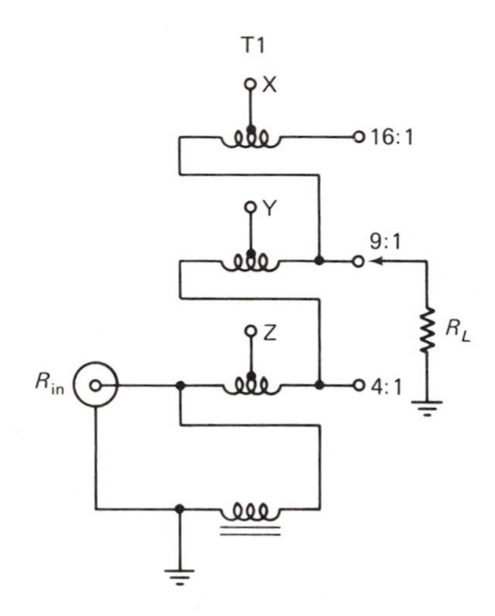

Figure 4.15 Hybrid type of broadband transformer for matching a variety of common impedances. This design was done by J. Sevick of Bell Laboratories. (M.F. "Doug" DeMaw, *Ferromagnetic-Core Design and Application Handbook*,©1981, p. 112. Reprinted by permission of Prentice-Hall, Inc., Englewood Cliffs, N.J.)

4.3.3 Printed-Circuit Transformers

Broadband and narrow-band transformers can be fabricated on copper-clad circuit board for use at HF and higher. In narrow-band applications the desired amount of inductance and the required transformations can be realized easily by forming the conductors through the etching process. If broadband operation is sought, flat slabs of ferrite can be placed on each side of the etched board and cemented in place. An advantage in this technique is that the compensating capacitors can also be etched on the circuit board. Production cost and time can be enhanced measurably by utilizing this method. Also, predictable results can be had from a production run once the prototype circuit board has been found acceptable. A simplified illustration of a broadband etched-circuit transformer is given in Fig. 4.16(a). Plates B and C contain the etched inductors of the transformer. In practice, the circuit-board foils would constitute several coil turns rather than the one-turn inductor of B and the two-turn example on board C. Double-sided pc board is used for plates B and C. This permits the large copper tabs of Fig. 4.16 to form compensating capacitors ($C1$ and $C2$) with the ground foil of each board.

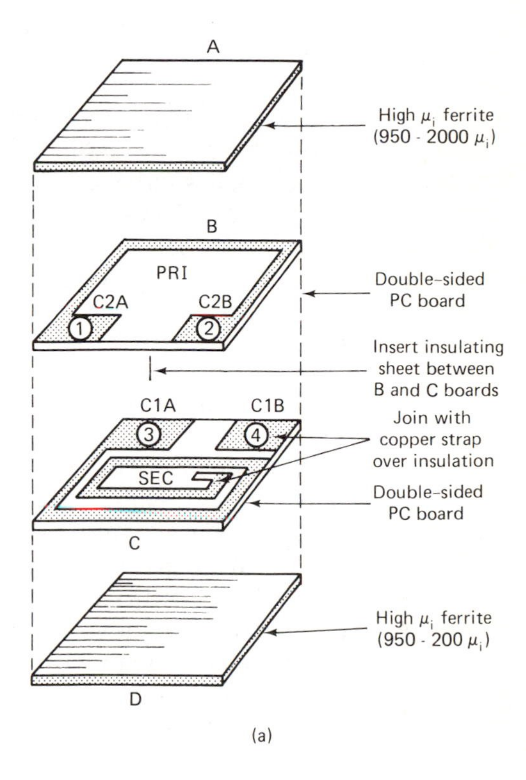

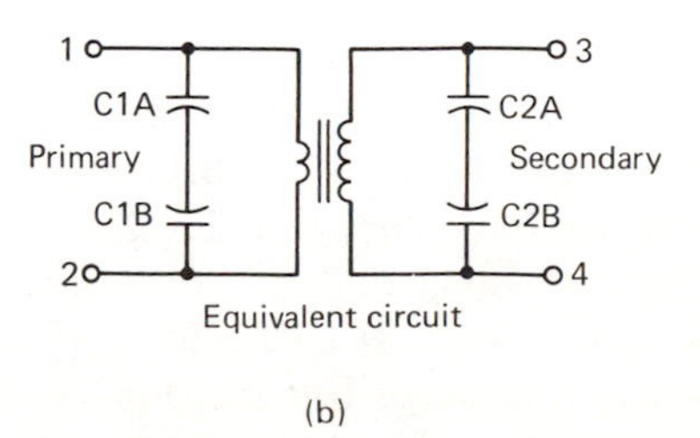

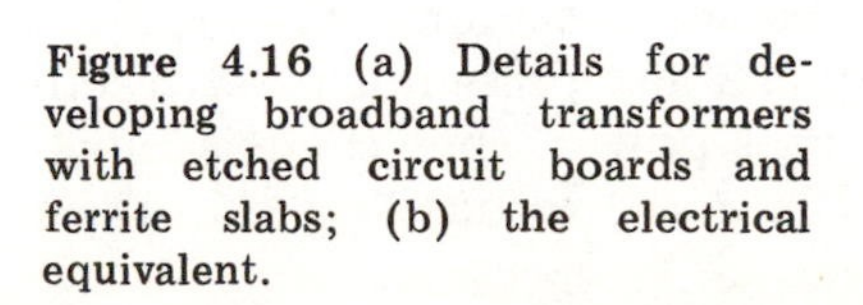
Figure 4.16 (a) Details for developing broadband transformers with etched circuit boards and ferrite slabs; (b) the electrical equivalent.

The inner coil end of plate C is joined to $C2B$ by means of a copper strap that has been soldered to the inductor and the tab. A thin strip of insulating material must be placed between the strap and the inductor to prevent short-circuiting. Finally, the end plates (A and D) are added to form a sandwich of all four pieces. Transformers that are built in this manner are not capable of handling large amounts of RF power, and are limited generally to use in the 10-W or lower power class. An increase in power capability could be realized if the pc boards used Teflon dielectric, and if Teflon sheets were placed between plates A, B, C, and D as insulating elements. The cost of such a technique would probably negate the virtues of this kind of transformer.

4.3.4 Hybrid Transformers

Hybrid transformers are commonly used as power splitters and combiners in solid-state amplifiers. Splitters are used to divide equally the power from a single source and apply it to a load pair, such as splitting a 4-W exciter output and applying the energy to the bases of the transistors in a pair of single-ended amplifiers. In this case there would be approximately 2 W of power delivered to each transistor base.

The combiner transformer is employed at the output of a pair of single-ended amplifiers to combine their powers and deliver the sum to a single load. We can see from this that hybrid transformers can be used for either purpose by simply reversing them to accommodate the application. The energy that is split or summed is of the same phase rather than being of 180° phase difference, as would be the situation if we used center-tapped transformers at the input and output of a push-pull amplifier. Isolation between the ports of a hybrid transformer is on the order of 30 to 40 dB for a properly constructed and applied unit, assuming we are operating from 1.5 to 30 MHz. The isolation effect allows an amplifier to continue operating even though one of the power sources may fail. Of course, failure of one of the amplifiers will result in reduced output power, but a constant load impedance will remain to protect the transistors from damage. Furthermore, the linearity of the amplifier will be maintained. Figure 4.17(a) illustrates how we would connect a hybrid combiner to a pair of amplifiers. $T1$ serves as a phase-reversal transformer for feeding V_{CC} to the collectors of the transistors. $T2$ is a combiner transformer that delivers the output of the amplifiers to a single-ended load. Both transformers are bifilar wound on ferrite toroid cores of suitable cross-sectional area. For illustrative purposes we have chosen a 15-Ω collector characteristic for each transistor. These two impedances are combined by $T2$ and present a 7.5-Ω terminal to the load. $R1$ of Fig. 4.17(a) absorbs power when there is imbalance in the amplifier output. It should be rated to handle 0.25 of the power output from the total system.

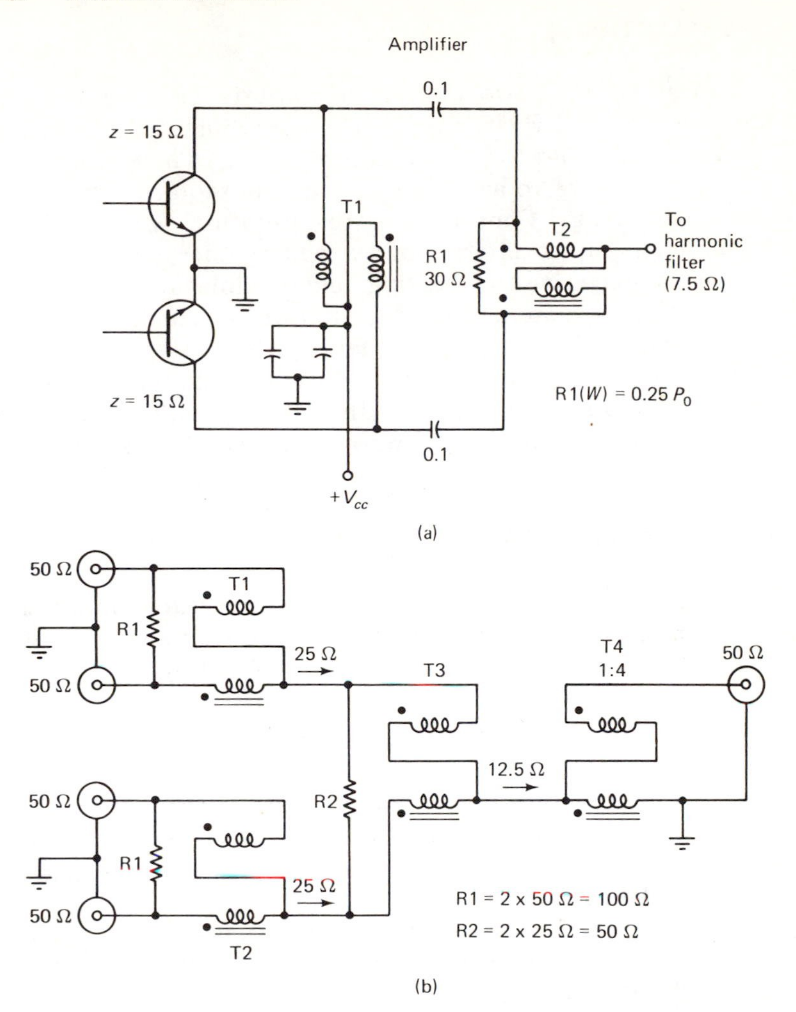

Figure 4.17 (a) A phase-inverting collector choke ($T1$) in combination with a hybrid combiner ($T2$); (b) a complex hybrid combiner using four broadband transformers. (M. F. "Doug" DeMaw, *Ferro-Magnetic-Core Design and Application Handbook.* © 1981, p. 108. Reprinted by permission of Prentice-Hall, Inc., Englewood Cliffs, N.J.)

The circuit of Fig. 4.17(b) is a complex hybrid combiner that contains four broadband toroidal transformers. $R1$ and $R2$ are balancing resistors to keep the VSWR at a low value, even if one of the amplifiers fails. Each resistor is twice the ohmic value of the driving source. As in the circuit of Fig. 4.17(a), each resistor should be rated for at least 0.25 of the system output power. The resistors are necessarily of the noninductive type. This general subject was treated in depth by H. Granberg in Motorola's *AN-749* application note.

4.4 CORE SELECTION

We are concerned with two matters of importance when selecting a magnetic core for an RF circuit. (1) The core must operate within its linear region at all times to prevent saturation. (2) There must be sufficient area on the core to accommodate all the required wire turns of the chosen wire gauge. Concerning core saturation, we want to avoid this phenomenon in the interest of preventing the generation of harmonic currents, excessive core heating, and possible core damage. When working with ferrite materials, hard saturation can change the μ_i of the core permanently. This is not the situation when using powdered-iron cores. But, in either instance, heating of the core material does change its permeability, so as a good rule of thumb we should always select a core that can handle considerably more power than we anticipate in a given circuit.

4.4.1 Operating Flux Limits

The flux density of a specified core is measured in gauss (G). Hence, the higher the developed flux, the more we approach potential saturation. Operational flux density (B_{op}) if defined by

$$B_{op} = \frac{V_{rms} \times 10^8}{4.44\, f N a_e} \quad \text{G}$$

where V_{rms} is the developed RF voltage, f is the operating frequency in hertz, N is the number of winding turns, and A_e is equivalent area of the magnetic path in square centimeters (cm^2). A_e is available in the manufacturers' data sheets.

If direct current is flowing through the transformer winding, we must use a variation of the basic equation to find the flux density. In this case we find B_{op} from

$$B_{op} = \frac{E_{rms} \times 10^8}{4.44\, f N A_e} + \frac{N I_{dc} A_L}{10 A_e} \quad \text{G}$$

where I_{dc} is the direct current in the winding and A_L is the manufacturer's inductance index for the core being used. A_L is usually given in the data sheets.

B_{op} can be found by other mathematical means as well, such as

$$B_{op} = \left[\frac{2500E}{f\sqrt{L V_e}}\right] \sqrt{\mu_e} \quad \text{G}$$

when f is in hertz, L is in henries, V_e is the effective core volume in cubic centimeters (cm^3) ($V_e = \ell_e \times A_e$), E is the rms excitation voltage, and μ_e is the effective permeability. We can also learn what B_{op} is from

$$B_{op} = k_1\sqrt{\mu_e}$$

where

$$k_1 = \frac{2500E}{f\sqrt{LV_e}}$$

From the foregoing we can relate permeability to B_{op} by noting that $\mu_e = B_{op}^2$ divided by k_1^2. Some engineers prefer to express E in the foregoing equations as the *peak* voltage rather than the rms value. By adopting E_{peak} as the term in our formula, we allow a reasonable margin of safety with respect to core saturation. The resultant increase in core size is minimal for most RF applications.

So that we may illustrate the determination of B_{op}, assume that we have a broadband transformer ($T1$ of Fig. 4.18) that transfers RF power from a 50-Ω driving source to the bases of two transistors in a push-pull amplifier. Our nominal operating frequency is 2.3 MHz and the driving power is 4 W. This will yield an E_{peak} across the primary of $T1$, the value of which will be 22.2 V. Since we are working in the medium-frequency spectrum rather than at audio, f in the B_{op} equation will be changed to megahertz. Also, E_{peak} will be multiplied times 10^2 rather than 10^8. From this evolves

$$B_{op} = \frac{E_{\text{pk}} \times 10^2}{4.44fNA_e} \quad \text{G}$$

Our next step in core selection is to study the manufacturer's data sheets and choose a trial-size core. Assume in this example we have settled for a Ferroxcube 76T188, type 4C4 toroid core, as it has a B_s ($B_{\text{saturation}}$) of 3000 G. We have determined from the manufacturer's A_L index for the core that N will consist of 14 turns of wire. Therefore,

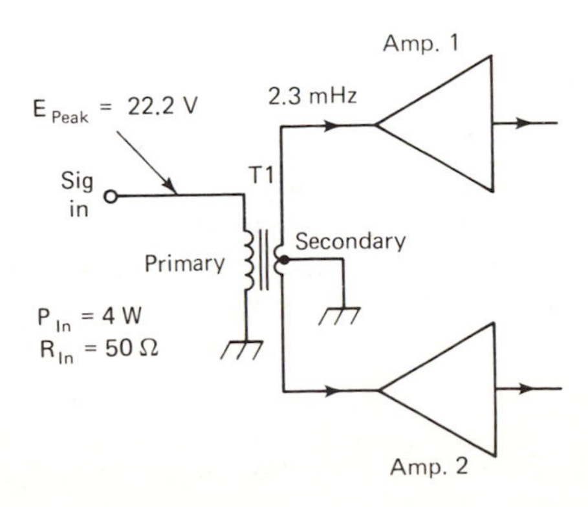

Figure 4.18 The B_{op} of a circuit using a magnetic core must be known when selecting the proper core material.

$$B_{op} = \frac{22.2 \times 100}{4.44 \times 2.3 \times 14 \times 0.133} = 116.75 \text{ G}$$

A_e for the core specified is 0.133, as given by the manufacturer. A B_{op} of 116.75 G is well below the rated B_s of the core, so the transformer will operate well within the linear range of the core. It is somewhat a trial-and-error procedure to select a suitable magnetic core, since the core must operate in its linear range as well as accommodate the required number of wire turns.

4.4.2 Number of Turns

For us to use the equations in Section 4.4.1, we need to know the value of N. This is determined easily by using the manufacturer's A_L factor for the core we have selected. For the Ferroxcube 76T188 used for the circuit of Fig. 4.18 we have a specified A_L of 70. The target inductance of the primary winding of $T1$ in Fig. 4.18 will be based on an X_L of four times the characteristic impedance of the driving source (50-Ω). Therefore, X_L will be 200 ohms. Using the standard equation for inductance we learn that

$$L = \frac{X_L}{\omega} = \frac{200}{6.28 \times 2.3} = 13.8\ \mu\text{H} \quad \text{or} \quad 0.0138 \text{ mH}$$

where ω equals $2\pi f$(MHz) and X_L is in ohms.

From this information we can calculate the required number of primary turns from

$$N = 1000\sqrt{\frac{L(\text{mH})}{A_L}} = 14 \text{ turns}$$

Some manufacturers provide A_L factors that relate to microhenries rather than millihenries. In situations of that kind the equation becomes

$$N = 100\sqrt{\frac{L(\mu\text{H})}{A_L}}$$

This variation of the formula is found most commonly in the literature for powdered-iron cores that are intended for use from 2 to 150 MHz.

There may be times when the A_L factor of a given core may not be easy to obtain. Should this happen, we can find the A_L by winding a few experimental turns on the test core, measuring the inductance, and applying the following equation:

$$A_L = \frac{L(\mu\text{H}) \times 10^4}{N^2}$$

where N is the number of turns on the core.

4.5 AMPLIFIER BIASING

Class C solid-state amplifiers are generally operated with zero bias, and draw no current other than that caused by leakage until excitation is applied. Amplifiers of this class are used for AM, CW, (radio teletype) RTTY, and FM operation, although linear amplifiers are suitable for the same modes and are sometimes utilized for amplifying signals of the foregoing kind. If bias is applied to class C solid-state amplifiers, the recommended method is to obtain it by means of emitter resistors rather than applying negative voltage to the base or exciting the amplifier across a shunt base resistor. The latter two techniques can cause damage to the transistors, which will be evidenced over time as beta degradation and subsequent failure of the transistors. Some designers employ emitter bias in the interest of enhancing the class C conduction angle to increase the amplifier efficiency.

Linear operation can be obtained by biasing the transistors in the class A, AB, or B modes. The class AB mode is the more common one for single-sideband transmissions. With most power transistors the three linear operating modes can be established over a forward bias range of 0.4 to 1 V. This makes it convenient to use a series resistor and clamping diode directly off the V_{cc} line for simple amplifiers. The forward bias will then be the barrier voltage (forward) of the diode unless a resistive divider is used to lower the bias further. High-power amplifiers of quality design usually contain a series pass transistor and an adjustable IC voltage regulator, which are operated from the V_{cc} line to provide the desired amplifier bias.

During class A operation the amplifier bias is adjusted to yield a collector idling current that is roughly 0.5 the peak current that occurs during peak signal conditions. This results in a 360° conduction angle.

For class AB operation the bias current can be approximated as I_C/h_{FE}, when I_C is the collector current based on a typical amplifier efficiency of 50%, and P_o (power output) referenced to this efficiency is used to determine the I_c from $2P_o/V_{cc}$. Factor h_{FE} (dc beta) can be taken from the transistor data sheet. Typically, it will be approximately 30. Based on this premise, and assuming that a collector current of, say, 10 A were flowing, the bias current would be 10/30 = 0.33 A. The conduction angle for class AB service will be somewhat greater than 180°.

If the transistors in an amplifier are biased for the class B mode, the conduction angle will be 180°. The bias voltage will be equal to the transistor V_{BE} (base-emitter voltage). No collector quiescent current will flow other than that of collector-emitter leakage.

Spectral displays of the effects of correct and incorrect biasing of linear amplifiers are shown in Fig. 4.19. Figure 4.19(a) shows the IMD products (third, fifth, and seventh) during a two-tone test to be 37 dB or greater below peak power. A level of - 30 dB or greater is considered

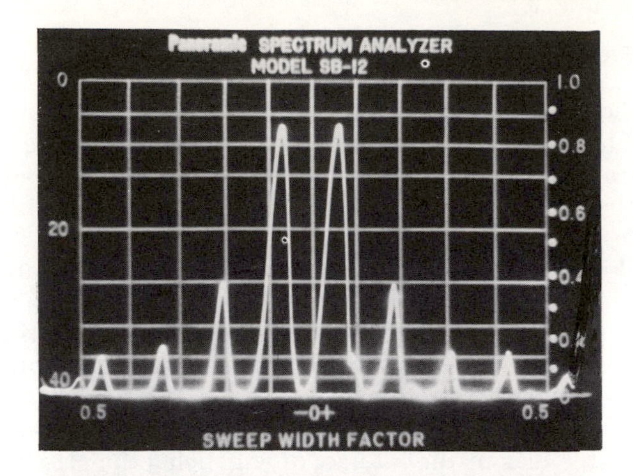

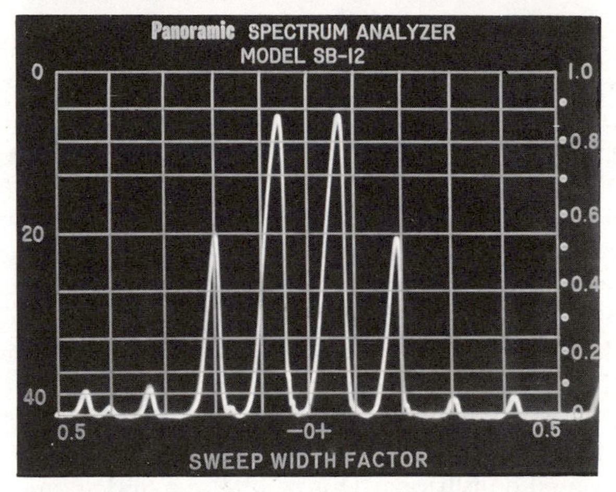

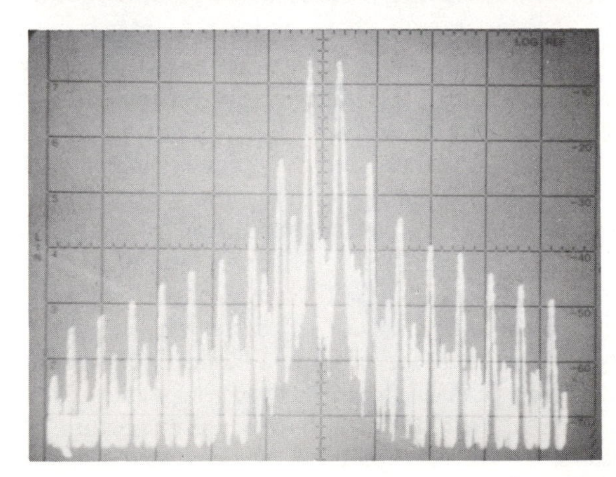

Figure 4.19 Spectrographs showing various levels of amplifier IMD during a two-tone test.

acceptable and easy to obtain. The display of Fig. 4.19(b) shows the third-order products down only 28 dB from peak power. This degradation was brought about by applying too little bias on the bases of the test amplifier. A worst-case condition is shown in Fig. 4.19(c), where severe IMD is seen in a VHF solid-state amplifier. The amplifier was intentionally operated class C and was made slightly unstable. Improper amplifier biasing and the subsequent distortion will cause the transmitter signal to be abnormally broad. A well-designed modern solid-state linear amplifier can yield IMD products that are 35 to 40 DB below peak power.

The actual amplifier IMD profile will be affected by the purity of the driving signal. A good rule of thumb is to have the IMD products in the excitation signal at least 6 dB lower than the target amount for the final amplifier. Harmonic energy from the driver should be well suppressed also. A level of at least -40 dB is suggested. If a heterodyne type of exciter is used ahead of the amplifier, all spurious products should be 50 dB or greater below peak output power from the exciter.

Figures 4.20 and 4.21 show methods for establishing linear-amplifier bias with diodes. The technique of Fig. 4.20 represents a somewhat brute-force biasing approach, in that it is limited by the barrier voltage

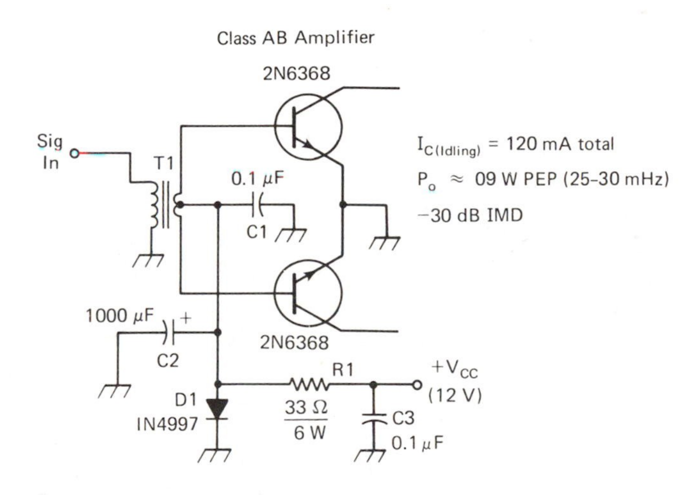

Figure 4.20 Simple method of applying forward bias to a linear amplifier transistor pair. *D*1 sets the bias at its barrier voltage.

of the diode used. For a germanium power diode the developed voltage would be on the order of 0.4 to 0.6 V, and with a silicon power diode it would be between 0.6 and 0.8 V, depending upon the characteristic of the diode chosen. However, this method is commonly used in low- and medium-power amplifiers to place amplifiers in the linear mode. It does not provide, however, for optimization of the amplifier IMD profile.

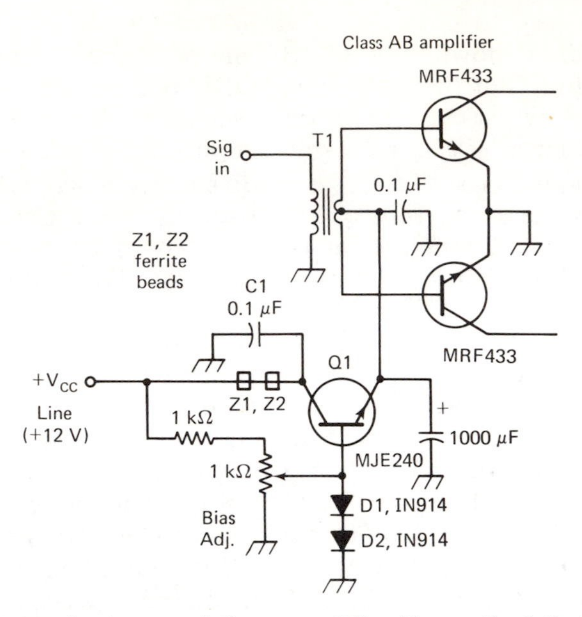

Figure 4.21 An improved linear-amplifier bias method that permits optimizing the V_{BB} voltage for lowest IMD.

Figure 4.21 shows an adaptation of a method suggested by Motorola to obtain variable bias by means of diode reference and an emitter follower ($D1$, $D2$, and $Q1$). Two diodes are used in this example. The second one has been added to compensate for an approximate 0.7-V barrier potential of the $Q1$ base-emitter junction. $R1$ provides a range of approximately 0.5 to 1 V. Small-signal diodes can be used in this circuit, whereas $D1$ of Fig. 4.20 must be able to handle a substantial amount of junction current. As mentioned earlier, IC types of voltage regulators can also be used to obtain variable bias from the V_{cc} line of the amplifier. In any event, the bias regulator needs to be well decoupled for RF from the collector circuit. This will prevent unwanted feedback from collector to base in the amplifier and will preclude the appearance of RF currents in the bias regulator, which could cause poor performance. $Z1$, $Z2$, and $C1$ of Fig. 4.21 serve as decoupling elements. Ferrite beads with a 950 μ_i are recommended at $Z1$ and $Z2$. $C1, C2, C3$, and $R1$ function as a decoupling network in Fig. 4.20. A more definitive treatment of decoupling methods is presented in the section on amplifier stability.

4.5.1 Thermal Considerations for Biasing

The foregoing discussion about biasing might make it seem that the entire procedure is pretty straightforward. The mechanics of developing the required bias are simple, indeed, but the matter of dc thermal stability can be a serious one. This is because the dc base-emitter voltage of the transistor drops as the junction temperature increases. Under this condition the application of a fixed-value forward base bias will cause the junction temperature to increase further by virtue of additional collector-current flow. These cascading events will eventually drive the stage into thermal runaway and ultimately destroy the transistor. A common practice has been to couple the bias diode to the case of the transistor or to the heat sink. The thermal bonding must be good to permit the diode characteristics under heating to track favorably with the changing characteristics of the transistor. Bias stabilization, though somewhat imperfect, can be achieved in this manner by taking advantage of the changing internal resistance of the diode with heating of its junction. Schemes have been devised whereby thermistors were coupled thermally to the transistor or its heat sink. Varying degreees of success have been realized during use of these simple techniques, but the greater the transistor h_{FE}, the less effective the protection measure becomes.

Communications Transistor Company devised a component called the BYISTOR in the 1970s. This device provides superb temperature tracking for dc stability, while providing a simple bias circuit. The BYISTOR is fabricated exactly as a power transistor for good temperature tracking. A silicon resistor is formed on the same substrate. The two components are packaged in stud, flange, or flangeless transistor case. This format permits mounting the BYISTOR on the same heat sink used by the transistors in the RF amplifier. As a result, the dc stability of the amplifier will be excellent and thermal runaway will not occur.

Figure 4.22(a) contains the circuit details of the bias device under discussion. None of the terminals is common to the case. $R1$ is the silicon resistor and $D1$ is the internal diode. Figure 4.22(b) shows the change in BYISTOR output voltage for various amounts of T_c (case temperature) in degrees Celsius. The curves suggest strongly that the BYISTOR be mounted as close to the amplifier transistors as practical. This will ensure optimum results in tracking. The recommended input current to the bias device ($U1$) is 350 mA. A variable resistor is shunted across the V_{out} terminals of the BYISTOR and adjusted for the desired quiescent collector current in the RF amplifier. The source impedance of the BYISTOR is a collective value set by the diode (0.3 Ω) and the resistor, $R1$ (0.7 Ω), resulting in an actual source impedance of approximately 1 Ω.

4.5.2 VMOS Power FET Biasing

We have an inherent trait working in our favor when we employ VMOS power FETs in RF linear amplifiers. This type of transistor has a

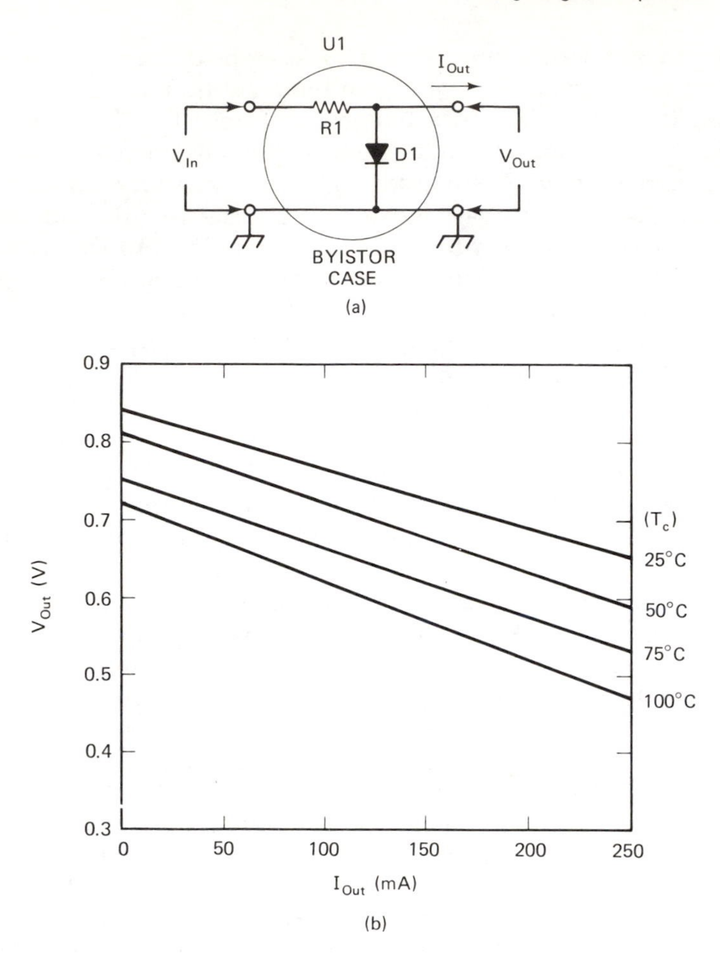

Figure 4.22 (a) Thermal compensation of the forward bias on a solid-state linear amplifier can be effected by using a BYISTOR device; (b) tracking curves.

linear transfer characteristic built in. Conversely, small-signal FETs exhibit an almost square-law response: the drain current is proportional to the square of the V_{gs}. However, the VMOS transistor, because of its short channel, yields a drain current that is linearly proportional to V_{gs}. Figure 4.23(a) shows a simple technique for obtaining the required forward gate bias for a VMOS linear amplifier. *D*1 is a Zener diode of appropriate value for the required bias. Some designers use only a simple resistive voltage divider to establish the operating bias. This is not difficult to do, since the gate draws very little current. Bias voltage can be obtained also by employing a series Zener-diode circuit.

The curve in Fig. 4.23(b) illustrates the excellent linearity of a typical VMOS power FET. We can observe that a forward gate bias of approximately 4 V will eliminate crossover distortion. This curve

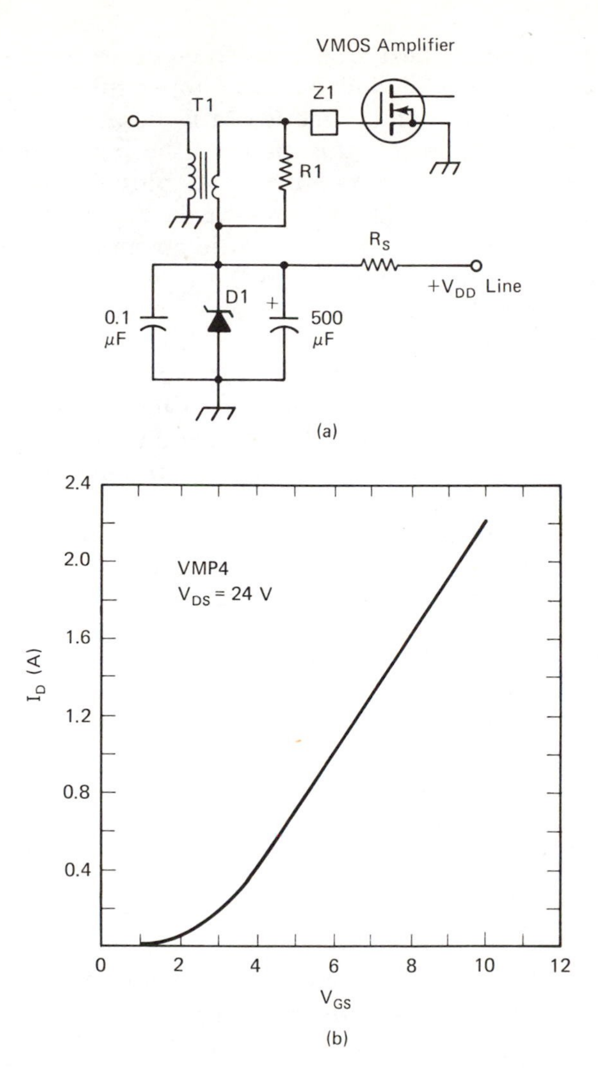

Figure 4.23 (a) Simple method of obtaining shunt-regulated bias for a VMOS power FET amplifier; (b) the curve shows the relationship between VMOS forward bias and drain current.

applies to almost all VMOS power FETs and should be useful in determining the bias requirements for linear operation. This curve compares favorably with a curve published by Siliconix, which was derived from an 80-μs, 1% duty cycle pulse test. $R1$ of Fig. 4.23(a) is a low-value resistor, chosen to permit a desired $T1$ transformation ratio from an existing source impedance, such as 50 Ω. $Z1$, a ferrite bead, is used as a parasitic suppressor. R_s is chosen to provide the required Zener-diode current.

4.6 HEAT SINKING

The matter of choosing a heat sink of adequate area for a given amplifier is a difficult assignment, indeed. No simple rules can be offered, and the subject is well beyond the scope of this book.

Many engineers approach this part of the design effort in true empirical fashion, striving for reasonable amplifier size (inclusive of heat sink) and adequate cooling to prevent damage to the transistors. One rule of thumb that works rather well in an empirical exercise is to apply full operating power and drive to the transistors once they have been installed on a heat sink. If, after sustained operation, the transistor cases are so warm that they are uncomfortable to the touch, the heat sink is too small. A transistor of the power type will be quite warm to the touch, as will the heat sink, but it should never feel hot. Admittedly, the technique is crude at best and is not recommended for commercial design work. It is usually adequate for breadboard and prototype applications.

Numerous factors influence the required size of a heat sink. One is ambient temperature. The intended environment for the composite equipment must be considered, such as tropical regions, land-mobile service, and sites where the ambient temperature is relatively constant. Another consideration is the equipment duty cycle, plus the amount of air circulation and venting common to the equipment cabinet. Perhaps the greatest matter of concern is the effectiveness of the thermal bond between the transistors and the heat sink. In this regard, a good grade of silicon heat-sink compound should be spread in a thin layer between the interfacing surfaces of the transistor and heat sink. Cooling fans are frequently used to blow air across the heat-sink surfaces as an aid to temperature reduction.

An excellent paper on this subject is found on page 17-12 of the *Motorola RF Data Manual*, Motorola, Inc., 1978. The reader is referred to that presentation for pertinent guidelines and equations for heat-sink selection. Additional information is offered therein concerning proper mounting techniques for various styles of power-transistor packages.

4.7 FEEDBACK AND GAIN LEVELING

Owing to the 6-dB gain slope per octave of bipolar transistors, some means for gain leveling must be included in a broadband amplifier. Objectively, we try to hold the gain variation from f(low) to f(high) within 2 dB. That is, if our broadband amplifier was designed to operate from 1.5 to 30 MHz, the gain should be relatively flat across that frequency spread.

Not only must we be concerned with the flatness of the gain, we need to ensure that the input VSWR of the amplifier is within a 2 : 1 ratio across the intended operating range. Commonly, the real part of the transistor input impedance (Z_{in}) is quite low, on the order of a fraction of an ohm to a few ohms. This resistive component is determined largely by the area of the power transistor by virtue of inverse proportion. Therefore, we can say that the resistive value is also inversely proportional to the power-output capability of the transistor we have selected.

The reactive part of Z_{in} is determined by the transistor input capacitance (C_{in}), which changes with frequency, and the parasitic inductive reactance of the transistor package. Therefore, the input reactance consists of a combination of C_{in} and X_L. At some discrete frequency the actual Z_{in} will be completely resistive, but it will be essentially capacitive at low frequencies and inductive at the higher frequencies. These X_L and X_C components will have a marked effect on the net value of Z_{in}, thereby dictating a need for some type of compensating network at the amplifier input in order to keep the VSWR within the target 2 : 1-or-lower range.

VSWR correction and gain leveling can be effected by including various R, C, and L components at the amplifier input. Since the real value of Z_{in} becomes higher as the operating frequency is lowered, the VSWR (and the amplifier IMD) will be affected noticeably. Compensating networks placed in parallel or series with the input will aid in obtaining a match at the lower frequencies. Also, some of the driving power will be lost (absorbed) in the compensating network, thereby helping to equalize the amplifier gain. The absorption will be the more pronounced in the lower portion of the intended signal spectrum. Since the reactive components of C and L are complex and difficult to define from the manufacturers' data sheets, some experimental effort will be required by the designer with regard to gain equalization and VSWR correction.

4.7.1 Input Compensation

Gain leveling can be accomplished by including the *RCL* components shown in Fig. 4.24(a). The equations and the data needed to work them are fairly complex. A thorough treatment of the subject can be found in the Amperex literature, *ECO-7114*, by Eindhoven. $R1$ and $L1$ of Fig. 4.24(a) are selected to provide frequency roll-off at the lower frequencies, but have a minor effect on high-frequency performance. $C1$ and $R2$ also affect the low-frequency response, with the greatest effect coming from $C1$. $R2$ will elevate the transistor Z_{in} with respect to the driving source, while $R1$ helps to lower it at low frequencies. A more complex input network is seen in Fig. 4.24(b), where

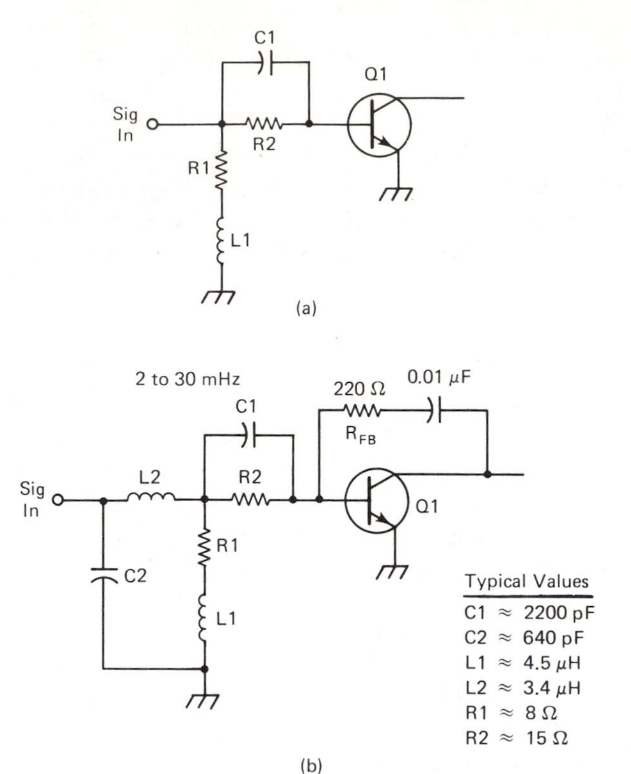

Figure 4.24 Two methods for providing frequency compensation at the input of an amplifier.

$L2$ and $C2$ have been added to provide additional compensating elements. Some typical component values are listed with the diagram to demonstrate what would be required for a 2- to 30-MHz broadband amplifier.

4.7.2 Shunt Feedback

Shunt feedback, also referred to as *negative feedback* and *collector-base feedback*, is a popular technique for effecting a trade-off between gain and bandwidth. The circuit of Fig. 4.24(b) contains an RC form of shunt feedback, which despite its simplicity is often adequate for the purpose.

A common system for obtaining feedback is illustrated in Fig. 4.25. $C1$ serves as a dc blocking capacitor. Therefore, it should have a very low X_c at the lowest operating frequency. With this circuit the low-frequency current gain of $Q1$ is roughly equal to the ratio of R_{FB} to Z_c. The feedback-resistance range is noted in step 1 of Fig. 4.25. $L1$ has been included in the interest of increasing the amplifier bandwidth.

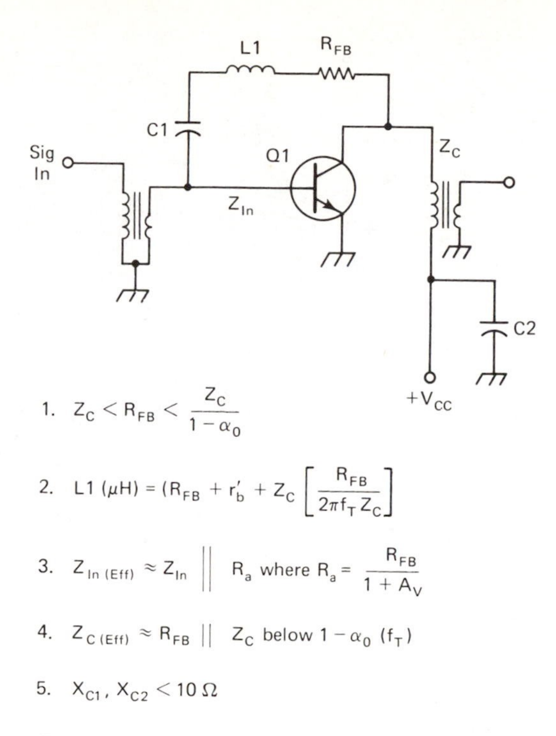

Figure 4.25 Example of shunt feedback and pertinent equations.

Without $L1$ the feedback resistance will be so low that it will load the input and output circuits. That unwanted condition will decrease the amplifier gain-bandwidth product. In essence, the inclusion of $L1$ enables us to realize the same gain-bandwidth product that would prevail if R_{FB} were not in the circuit. The feedback resistance tends to vanish with $L1$ present at frequencies above f_T, thus preserving the amplifier gain at the higher frequencies. For most RF power amplifiers the value of R_{FB} will lie in a range of 100 to 1000 Ω.

4.7.3 Degenerative Feedback

We can utilize still another method for gain leveling and bandwidth enhancement. The technique is known as emitter degeneration or degenerative feedback. It requires a resistor-capacitor combination, such as that shown in Fig. 4.26. R_E is a low-value resistance and C_E is tailored to yield the desired high-frequency gain while reducing the gain at low frequencies. C_E, therefore, reduces the degenerative feedback at the higher frequencies, while R_E increases it at the lower frequencies. The gain disparity in a 2- to 30-MHz broadband amplifier can be greater than 15 dB from the high to the low end of the chosen spectrum; it is

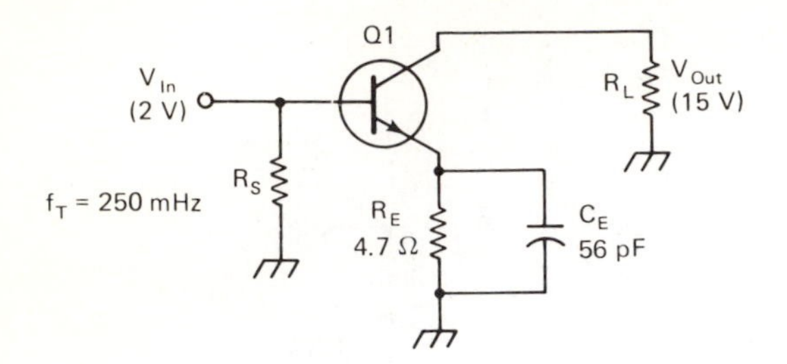

1. $C_E \approx \dfrac{1}{15 f_T\ R_E} = \dfrac{1}{15 \times 250 \times 4.7} = 0.00056\ \mu F$

2. $BW \approx \dfrac{f_T}{h_{fe}} = \dfrac{f_T\ (1 - \alpha_0)}{\alpha_0}$

3. Choose $R_S + r_b' > R_E$ for best BW

4. $A_V = 20 \log_{10} \dfrac{V_O}{V_{In}}$ dB = 8.75 dB

Figure 4.26 Degenerative feedback can be employed by means of R_E and C_E in this circuit. This type of bias is often used in combination with shunt feedback in broadband amplifiers.

imperative to employ gain leveling for this reason. Increased bandwidth will be possible when $R_S + r_b'$ is greater than R_E.

Degenerative feedback is avoided by many designers of high-power solid-state amplifiers in preference to shunt feedback. This is because the resistor and capacitor leads present unwanted inductive reactances in the emitter circuit, and this can cause instability. Most resistor and capacitor pigtails exhibit inductances up to 1 μH, which represents a substantial amount of X_L, when r_e' is low.

This form of feedback can work advantageously in combination with shunt feedback, since the latter lowers the amplifier input impedance, and degenerative feedback increases the input impedance. In combination, the two feedback methods tend to oppose one another in terms of Z_{in}, thus nullifying the effects of feedback on Z_{in}.

An example of a practical 2- to 30-MHz linear amplifier is offered in Fig. 4.27 to illustrate the feedback and gain-leveling format found in many amplifiers designed by Motorola. Collector-base blocking capacitors can be avoided by including a third winding ($L1$) on phase-reversal transformer $T3$, as shown. Keeping blocking capacitors out of the very low impedance collector circuit is beneficial because of the stray X_L components. Power losses can be reduced substantially by avoiding collector blocking capacitors between $T3$ and the following 16 : 1 matching transformer. The losses are caused by the high collector currents (IR losses). $C1$, $C4$, and $C5$ are compensating capacitors for $T1$ and $T3$, while $C2$, $C3$, $R1$, and $R2$ comprise the gain-leveling networks for the amplifier. The impedance ratios of $T1$ and output matching transformer (not shown) are noted in the diagram. P_o for this amplifier is 140 W,

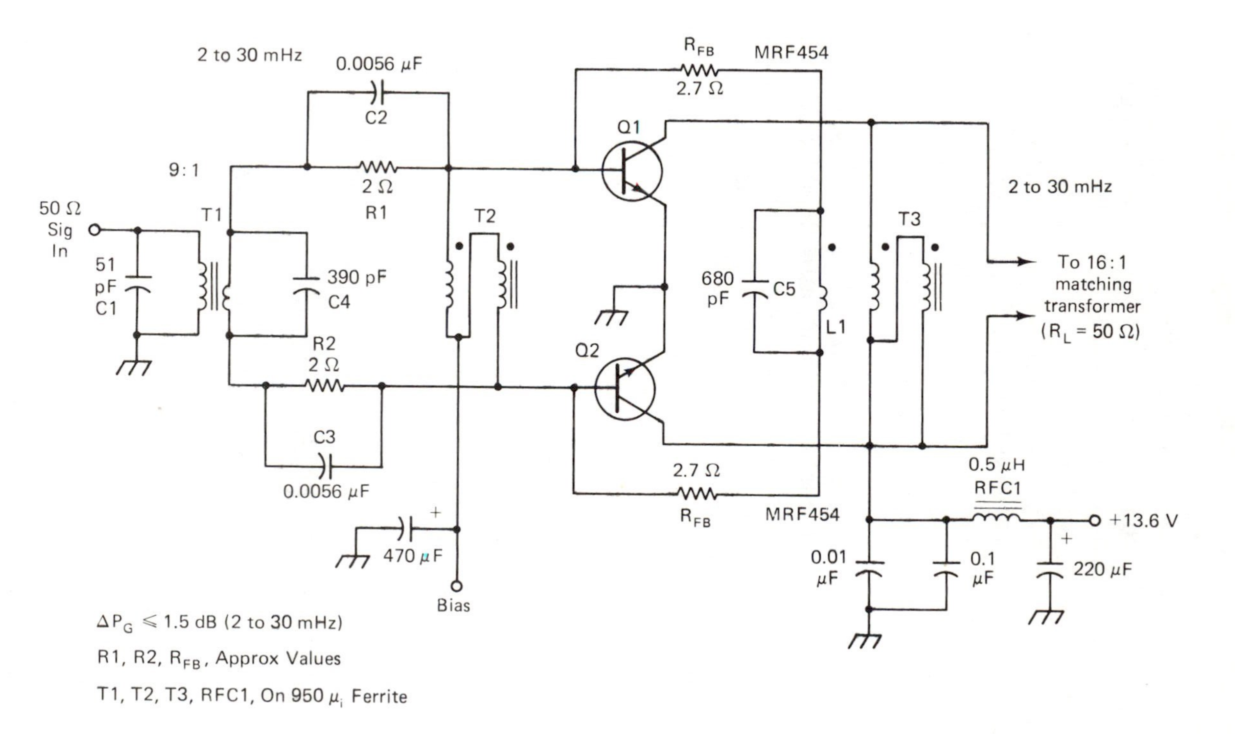

Figure 4.27 A practical circuit patterned after one that was suggested in the *Motorola RF Data Manual*, 1978. This circuit demonstrates the use of shunt feedback, with the feedback pick-off point being $L1$ of $T3$. This avoids direct connection to the collectors of $Q1$ and $Q2$. Feedback-component values will vary with the amplifier layout and the transistors used.

yielding a collector-to-collector impedance of approximately 2.6 Ω. This calls for a 19 : 1 impedance ratio to match a 50-Ω antenna load. The standard 16 : 1 transformer provides an integer that is close enough to allow satisfactory power transfer to a 50-Ω load.

A more elaborate broadband HF amplifier is described schematically in Fig. 4.28. It was inspired by design theory found in the Motorola literature. It is presented mainly to show applications of shunt and degenerative feedback, along with typical gain distribution from 0.1 W of input power to 100 W of P_o. $T1$-$T3$ are wound on 950 μ_i ferrite cores, as are the decoupling chokes, $RFC1$-$RFC5$, inclusive. $C1$-$C4$ are ceramic-chip capacitors that are soldered to the amplifier pc-board foils. This will minimize stray inductance and improve the amplifier performance. Two of the bias methods described earlier in the chapter are used in the circuit of Fig. 4.28.

It can be seen that extensive decoupling of the dc lines has been assured by addition of the RF chokes and their associated bypass capacitors. This practice is particularly important in power amplifiers, since high RF currents will otherwise flow along the V_{cc} line from one stage to another. Bypassing has been included for VHF down through the VLF part of the spectrum. Resistive loading has been applied to the bases of $Q2/Q3$ and $Q4/Q5$ as an aid to stability.

$R1$ can be chosen to limit the high-frequency response if that is desired. It will serve also to elevate the Z_{in} of $Q1$ and will function as a VHF parasitic resistor to discourage that type of instability. The 1000-Ω resistor across the primary of $T1$ was added to further broaden the response of amplifier $Q1$ by lowering the Q of the transformer winding. The characteristic Z_c of $Q1$ will be approximately 200 Ω using $Z_c = V_{cc}^2/2P_o$. Therefore, the 1000-Ω swamping resistor will have a minor effect on the transistor Z_c.

4.8 THE MATTER OF STABILITY

We established earlier in the chapter that the use of feedback helps to ensure stability in solid-state amplifiers. In an ideal amplifier there will be no apparent instability with or without the amplifier input and output ports terminated. In other words, the *open-loop stability* will be excellent. A properly stabilized amplifier will exhibit what is known as *unconditional stability*, and that should be our design objective. There are a number of measures we can take to provide high stability, so let's examine some common preventive steps.

Figure 4.29 shows a simple, single-ended class C amplifier that has no intentional feedback. We will assume in this example that we desire a single-frequency amplifier which operates on 2.3 MHz. That being the case, it would be rather pointless to incorporate shunt or degenerative

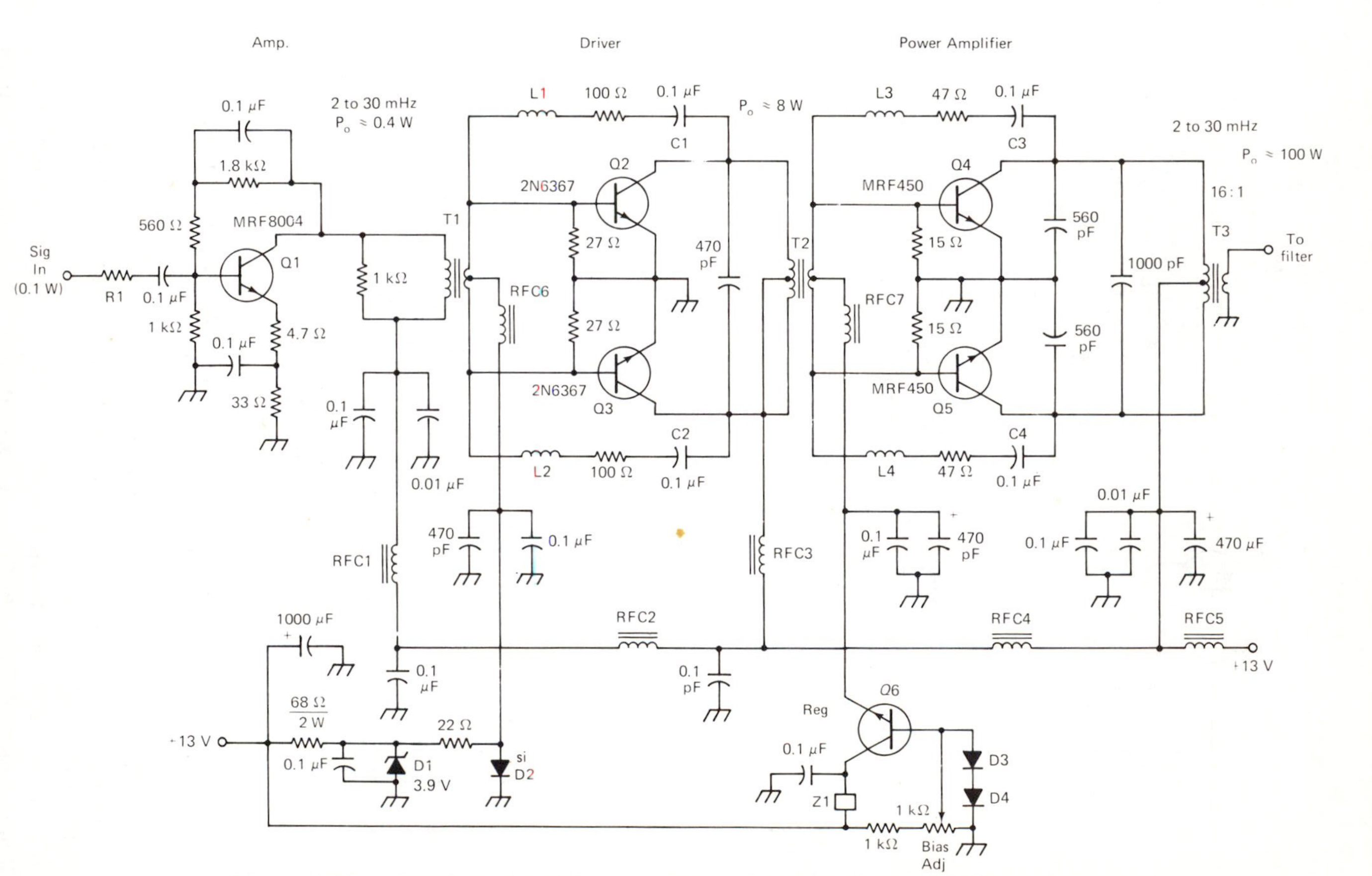

Figure 4.28 Practical circuit for a multistage broadband 2- to 30-MHz power amplifier. It illustrates shunt feedback and stability measures that are typical in commercial design. Overall gain for this circuit is 30dB, inclusive of transformer losses.

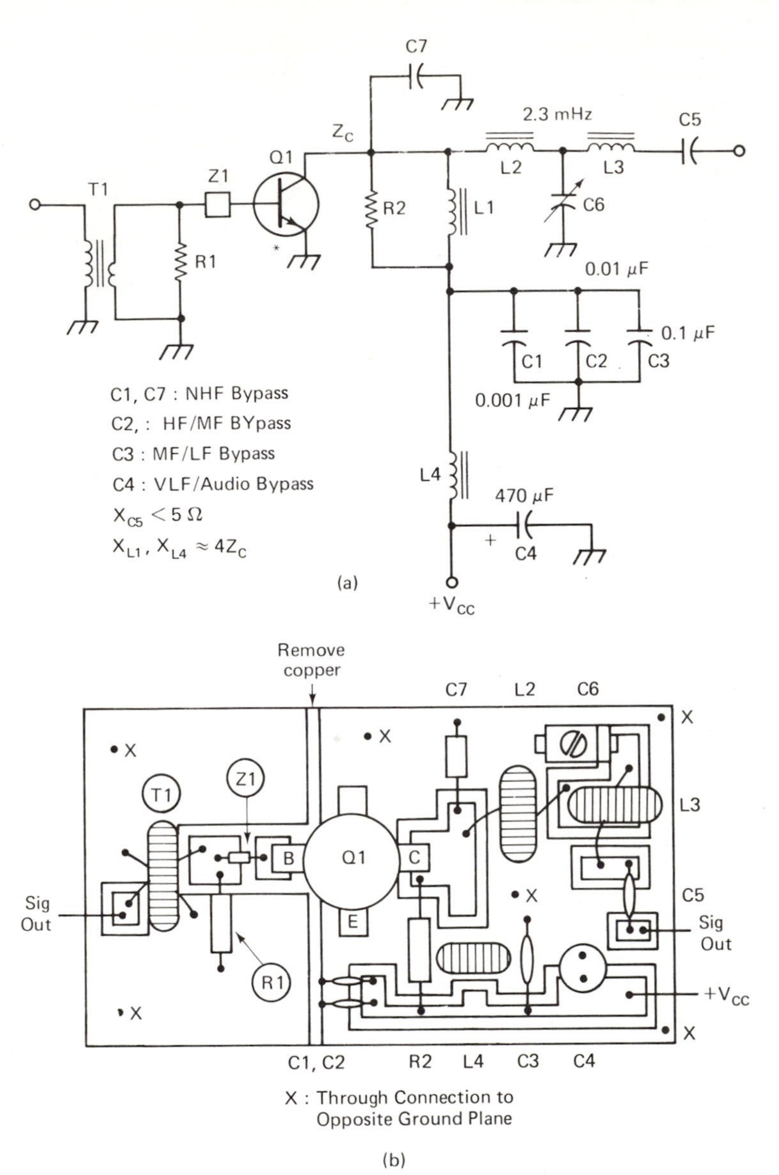

Figure 4.29 (a) Common techniques for ensuring amplifier stability; (b) a suggested pc-board layout for good stability.

feedback elements. By avoiding feedback we have left the door open to self-oscillations, and if the gain of $Q1$ is high enough, we are likely to encounter stability problems.

$R1$ of Fig. 4.29 can be included in the base circuit to lower the Q of $T1$ and hence discourage it from forming part of a broadly resonant tuned-base/tuned-collector oscillator. $R1$ is generally on the order of 5 to 22 Ω and represents for the most part an experimental choice. A good rule is to use only that amount of resistance that will aid stability. Some driving power will be dissipated in $R1$, so it should be the highest value practical. $Z1$ is a ferrite bead that is located as close to the base of $Q1$ (near the transistor body) as possible. It will discourage VHF and UHF parasitic oscillations. The asterisk seen near the emitter of $Q1$ in Fig. 4.29 indicates that the emitter to ground connection must be as short as possible to prevent unwanted inductance.

$C7$ is included as a low reactance at VHF and UHF. This capacitor will not only bypass VHF harmonic currents, but will aid in damping VHF and UHF oscillations. Its value of X_c should be high at the desired operating frequency to prevent power loss and matching problems at 2.3 MHz.

The use of $R2$ across collector choke $L1$ is somewhat a brute-force stabilization measure. This approach is not recommended except in very stubborn cases of instability. In effect, it swamps the Q of $L1$ in a fashion similar to $R1$ across the secondary winding of $T1$. $R2$ will dissipate RF power in accordance with the voltage swing at the collector. Assume that we had an rms collector voltage of 20. If $R2$ were 270 Ω we would dissipate 1.48 W in the resistor. The better solution would be to make $L1$ a low-Q inductor by selecting a higher μ_i core material that is rated for optimum Q at some lower frequency. For most applications in the medium- and high-frequency range a 950 μ_i core will be satisfactory. A 125 μ_i core might yield too high a choke Q if instability were likely.

V_{cc} decoupling should be done with all frequencies below f_T in mind. $C1$-$C4$, inclusive, were chosen with that simple rule in mind. We must be cautious also about the lead lengths of the capacitors, since even short pigtails introduce inductive reactances that can ruin the bypassing capabilities of the capacitors. In an ideal example there would be no capacitor leads whatsoever. We can approach that objective by using ceramic chip capacitors at those circuit points. The same is true of blocking capacitor $C5$. $L4$ serves as part of the decoupling network. Both $L1$ and $L4$ should be able to carry the collector current of $Q1$ without undue IR voltage drop and heating. Decoupling networks of the type seen in Fig. 4.29 assist in preventing RF currents from migrating along the V_{cc} supply line to other stages of a transmitter: stage-to-stage isolation is important in the prevention of instability.

We must also avoid physical layouts that encourage instability. That is, the input and output halves of an amplifier require reasonable isolation if we are to prevent self-oscillation. A suggested layout for the circuit of Fig. 4.29(a) is offered in Figure 4.29(b). The employment of double-sided circuit board discourages VHF and UHF instability, because the circuit-board pads, in combination with the board dielectric and the copper ground plane (opposite the etched side), form bypass capacitors that are effective at VHF and higher. The large copper ground surfaces also help to distribute circulating currents (ground loops) more uniformly than would be true of single, narrow conductor strips. This also helps ensure stability.

A strip has been etched from the top ground foil of the circuit board just to the left of the *Q*1 body. This prevents collector currents from following the ground plane (top) to the input of the amplifier. Some designers believe that an even better approach is to return all the input-component grounds to the opposite side of the pc board, while connecting the collector-circuit ground returns to the top ground foil on the board. If that method is physically practical, it should be adopted. The areas marked with an X in Fig. 4.29(b) indicate suggested points at which a through-connection is used to join the two ground planes. A short piece of bus wire is suitable for the purpose, with solder applied on both ends to join them to the copper foil. The heat sink is mounted on the side of the pc board that is not shown in Fig. 4.29(b).

The pictorial representation in Fig. 4.29(b) contains component leads that are substantially longer than they should be in the interest of amplifier stability. The component pigtails were made long to clarify the illustration. Furthermore, *C*1, *C*2, *C*3, and *C*5 are illustrated as disk ceramic capacitors, whereas the rectangular chip type of capacitor would be much better toward assuring stability.

4.9 TRANSISTOR PROTECTION

Despite modern design trends, the power transistor is still a fragile device under certain electrical conditions. The basic causes of damage or destruction are excessive voltage or current. Immediate destruction will almost always result if the applied operating voltage is of the wrong polarity. This is especially true of power transistors. Protection against cross-polarized voltages from an outboard power supply can be ensured by placing a power silicon diode between the power supply and the transistorized equipment it operates. The diode barrier voltage must be taken into account when designing the power supply, thus compensating for the 0.7-V drop across the diode junction. If large amounts of current are taken through the diode junction, voltage regulation to the load (equipment package) will suffer somewhat if continuous operation does not take place. This will result from the forward resistance of the diode. The diode, when placed in series with the positive supply line, will pass positive voltage but will prevent the passage of negative voltage if the supply leads are accidentally reversed.

4.9.1 Overvoltage Protection

Little is said in the manufacturers' literature about damage to transistors from excessive terminal voltages. The emphasis seems to be on the catastrophic effects of high current and dangerous junction temperatures. The VMOS power FET is especially prone to damage from excessive gate current and voltage. The same is true of too high a V_{DS} condition, even momentarily. Unprotected VMOS FETs can be destroyed immediately if the V_{GS} or V_{DS} ratings (maximum) are exceeded. For most VMOS transistors the specified maximum gate-source voltage is 30, p-p. Voltages beyond that amount can cause excessive gate current or may puncture the thin layer of silicon oxide that serves as gate insulation.

One technique we can adopt for protecting the gate of a VMOS FET is shown in Fig. 4.30(a). Zener diodes are connected in series with small-signal silicon diodes to clip both the positive and negative signal peaks. The effects of added shunt capacitance are reduced by including diodes $D5$ and $D6$ as shown. Clipping will occur at approximately ±15 V with this circuit. This preventive measure is useful when transient voltage peaks originate from switching action in circuits ahead of $Q1$, such as those we might generate by employing a mechanical band switch in the low-level stages of the transmitter.

Self-oscillations of the type discussed in the earlier part of this chapter can also cause excessive V_{GS} and V_{DS} voltages. Therefore, stability is highly important to device longevity. In this regard, the addition of the Zener-diode clamps will tend to damp VHF oscillations by virtue of their shunt capacitances.

Protection can be given to the drain circuit of a VMOS FET amplifier by employing Zener diodes from drain to ground, as seen in Fig. 4.30(a). $D3$ and $D4$ must be capable of handling voltages in excess of the maximum drain-source signal swing. This is on the order of $2V_{DD}$. Therefore, if we had a 24-V drain supply, $D3$ and $D4$ should be rated at slightly more than 48 V. A pair of 56-V Zener diodes would be entirely suitable. Arranged as shown, $D3$ and $D4$ will also clamp on positive and negative V_{DD} spikes that might originate in the power supply. No adverse operating effects should be noted up to at least 60 MHz by adding $D3$ and $D4$ to the drain circuit. For most applications we can use 1-W Zener diodes at the amplifier output. Diodes $D1$ and $D2$ can be 400-mW units.

Figure 4.30(b) shows the use of a Zener diode in the collector circuit of a bipolar transistor. There is no need to include a clamping diode from base to ground, since the base-emitter junction of $Q1$ serves that purpose.

4.9.2 VSWR Protection

VSWR protection (mismatch protection) can be applied to a bipolar-transistor amplifier by sampling the reflected power from the load, rectifying it, and using the resultant voltage to reverse-bias one of the low-level stages in the driver section of the transmitter. The general

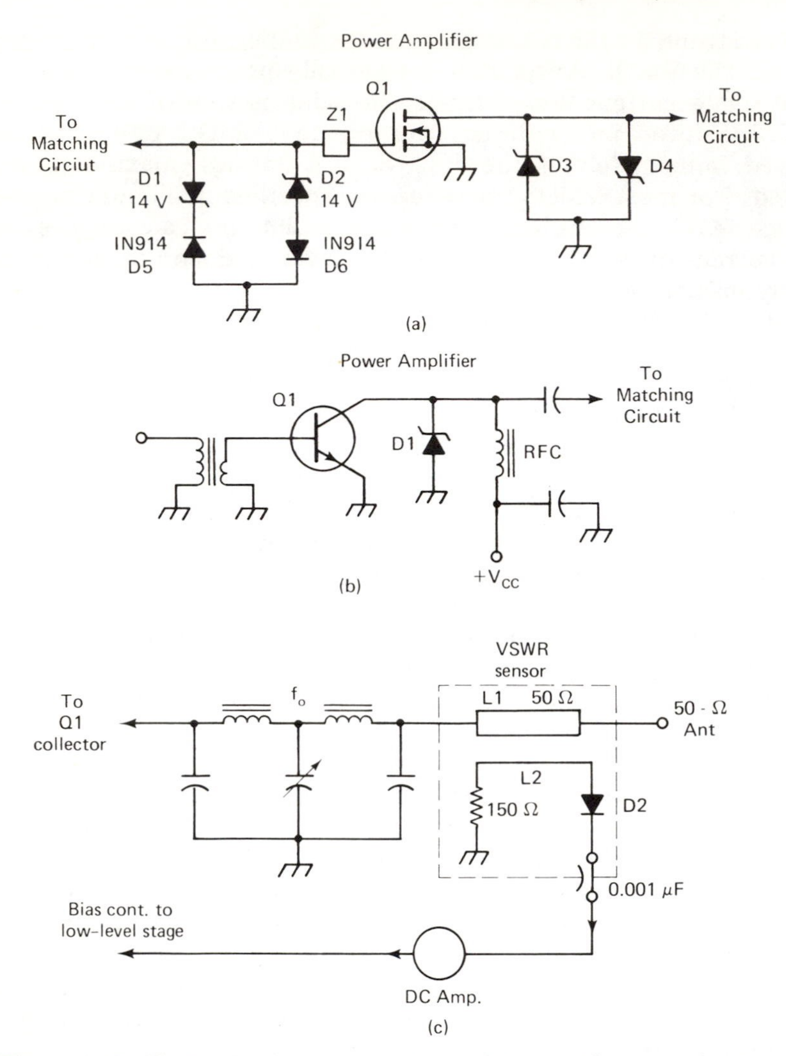

Figure 4.30 Various techniques can be used to provide protection to power transistors: (a) Zener diodes are used to prevent damage from voltage excesses in a VMOS amplifier; (b) a clamp diode is connected to the collector of a bipolar transistor for the same purpose; (c) VSWR protection is ensured by using a sensor circuit.

technique for achieving this is shown in Fig. 4.30(c). A short section of 50-Ω line is used between the output filter and the antenna. *L*1 and the shield compartment are arranged to yield a 50-Ω line characteristic. *L*2 samples the reflected component on the line. The energy flowing into the 150-Ω termination is rectified by means of *D*2 and then routed to a dc amplifier that yields the reverse-bias range required. The greater the VSWR, the lower is the output power or voltage from the low-level stage being controlled. Needless to say, excellent decoupling of the control line after $D2$ is necessary in the prevention of unwanted stage-to-stage feedback. A toroidal current-sampling VSWR sensor can be used in place of the circuit shown in Fig. 4.30(c). The limitation of the 50-Ω line section shown is its *frequency sensitivity*. That is, the higher the operating frequency for a given line length at $L2$, the greater is the output voltage from $D2$. It is better, therefore, to use a broadband toroidal sensor circuit for amplifiers that cover a wide frequency range, such as 2 to 30 MHz. The circuit shown in Fig. 4.30(c) is well suited to use in single-band HF or VHF amplifiers. VSWR-shutdown circuits of this general type will prevent damage to the PA transistors when the antenna system does not present the intended load impedance.

A practical circuit for a toroidal type of VSWR bridge is given in Fig. 4.31. The forward and reflected components on the transmission line

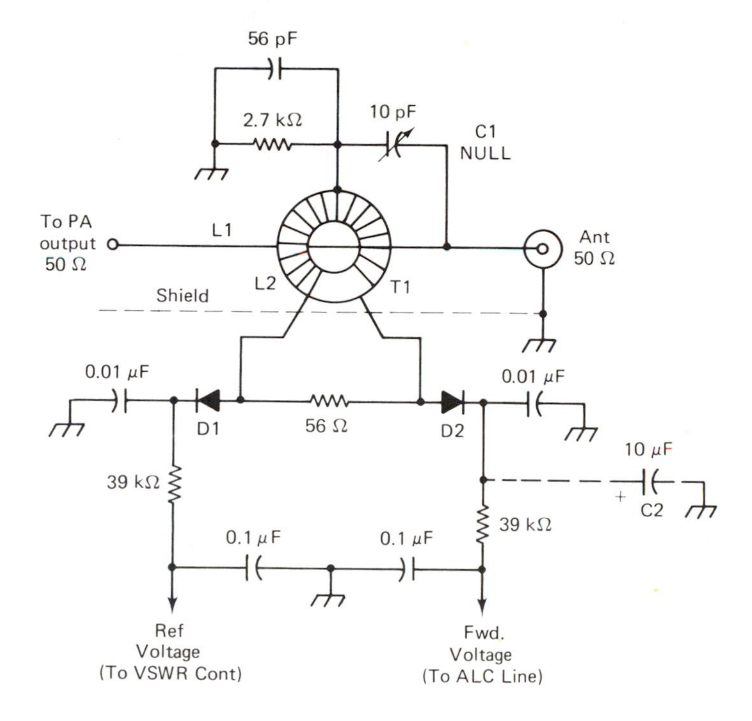

Figure 4.31 Practical circuit of a current-sensor, broadband VSWR sensor.

are sampled by $T1$. The center conductor ($L1$) of the transmission line constitutes a one-turn primary for $T1$. $L2$ is the secondary winding of the transformer. For operation from 2 to 30 MHz the $L2$ winding can be 60 turns of number 30 enameled wire on a 0.68-in.-diameter powdered-iron toroid core of Micrometals Corporation type 2 material. The component values specified in Fig. 4.31 are suitable for powers up to 1000 W. $C1$ is used to null the bridge with power present and with a

50-Ω source and 50-Ω resistive load. $D1$ and $D2$ rectify the sampled ac components to provide direct current for driving a 100-μA dc meter that can be switched to read forward and reflected power. $C2$ can be added to provide a time constant (forward-reading mode) that will help to read peak power for SSB or CW operation.

The forward voltage can be used to operate an automatic limiting control (ALC) in the low-level stages of the transmitter. This will help to preserve the amplifier linearity by limiting the peak drive to the power amplifier. Ideally, the ALC sampling point is situated at the input to the final amplifier, thereby controlling the drive power *before* distortion occurs. However, some manufactured amplifiers have the ALC take-off at the amplifier output.

The reflected voltage can be used as a VSWR-shutdown measure, as detailed earlier in this section. $L1$ should be kept as short as practical, with shielding between $T1$ and the lower part of the circuit, as indicated by dashed lines. $L1$ and the shield compartment walls can be structured to create a 50-Ω line impedance. This will not be necessary if microstrip techniques are used. Also, the bridge circuit may be more completely nulled at the upper frequency range if a Faraday shield is inserted between $L1$ and $L2$. It can consist of a slotted section of brass or copper tubing (slotted to prevent a shorted-turn effect).

4.10 THE MATTER OF SPECTRAL PURITY

A well-engineered transmitter will deliver a reasonably clean output signal to the load at all times. Harmonics, spurious responses, and wide-band noise need to be suppressed in accordance with government regulations to prevent interference to other communications services.

Wide-band noise, as shown in the spectrograph of Fig. 4.32(a) will be the greatest when the tuned circuit Q is low. The illustration of Fig. 4.32(a) was recorded with a Hewlett-Packard spectrum analyzer at an operating frequency of 5 MHz. It shows the noise output of a commercial transmitter local oscillator. The vertical divisions are 10 dB each, and the horizontal divisions are 2 kHz each. Much better noise profiles than the profile of Fig. 4.32(a) are possible through the use of high-Q resonators.

Another influence on spectral purity is the matter of spurious responses. In an ideal spectrum we would find only the desired signal information. In practice, this does not happen. We can, however, minimize the levels of spurious energy to an acceptable and legal value by means of good engineering practice. One type of spurious output is shown in Fig. 4.32(b). The highest peak on the spectrograph represents the transmitter carrier frequency. The lower-level, close-in spurs are 43

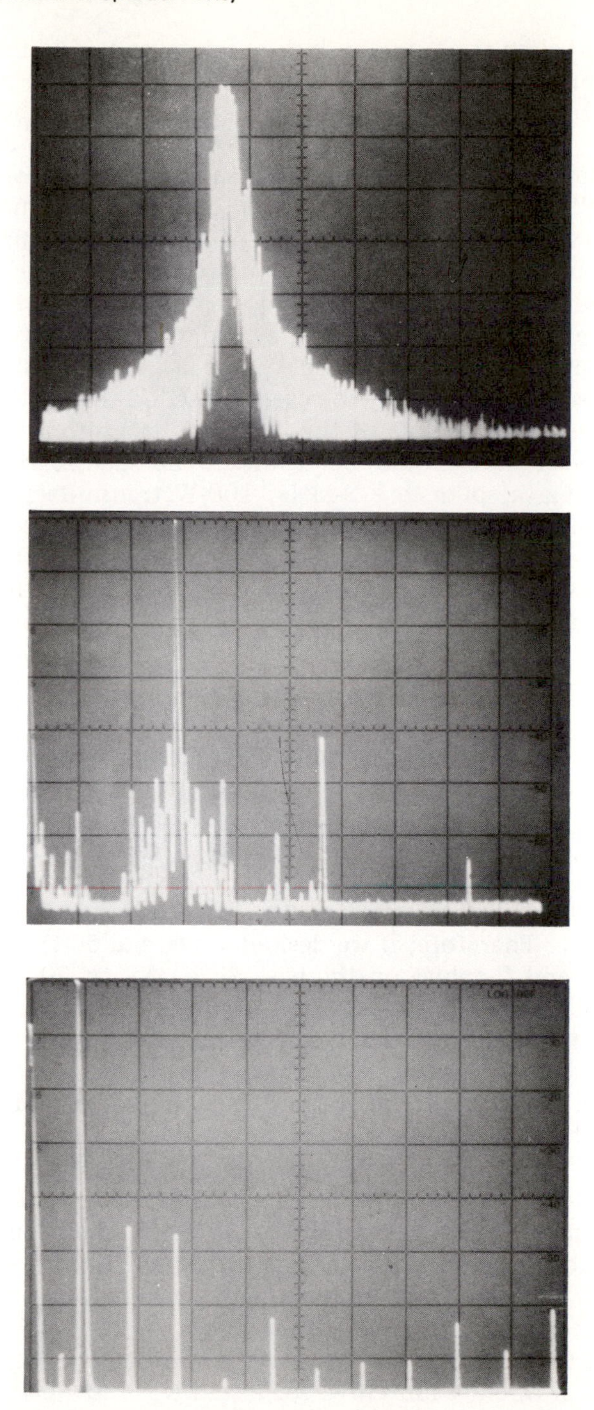

Figure 4.32 Spectrographs showing (a) wide-band noise, (b) in-band and out-of-band spurious responses, and (c) harmonics.

dB or more below peak power, but could cause undue interference to other communications service nearby. The unwanted responses shown are caused by mixing products in the early stages of the transmitter. Other spurs are the result of parasitic oscillations. The large response near the center of the spectrograph is the second harmonic (−42 dB), and the small response at the far right is the third harmonic (−64 dB). It can be seen that there are a number of spurious responses below the carrier frequency (far left) also. A spectral condition like that of Fig. 4.32(b) is unwarranted and is the result of inferior circuit design. A clean transmitter will show only the carrier and the harmonics during spectral analysis. Such an example is seen in Fig. 4.32(c). The harmonics are seen to the right of the carrier (second through the eleventh). The response at the far left of the picture is the zero reference marker of the analyzer and should be ignored. The spectrograph was made while analyzing a solid-state 2-MHz, 100-W transmitter. Although all the harmonics are within FCC compliance, the high-order harmonics are not suppressed as well as they could be. This suggests poor filter design and indicates inferior ultimate attenuation. Generally, this is a matter of filter layout and shielding.

4.10.1 Filtering

Close-in spurious energy cannot be removed by means of filtering, but harmonics and some far-out spurious products can be attenuated effectively through the use of low-pass or bandpass filters. One of the most common and simple types of low-pass filter used in transmitter design is the *half-wave* filter. It consists of two pi sections of L and C in cascade. This network is shown in Fig. 4.33(a). It is designed for a loaded Q of 1. Therefore, if we desired a bilateral 50-Ω filter we would select the L and C values on the basis of an X_L of 50. The input and output capacitors would have an X_C of 50 also, but the center capacitor would require an X_C of 25, since we would be joining two pi sections that had end reactances of 50 Ω. A sample design is provided in Fig. 4.33(a), based on an operating frequency of 8.5 MHz. The filter cutoff frequency will be somewhat higher than f_o, and a factor of 1.15 will be satisfactory for most applications. This is done to prevent in-band attenuation by the filter. The component values listed in the schematic illustration are computed ones. Since the resultant numbers are not suitable for standard-value capacitors, the nearest standard value can be used, and there will be minor impairment to the filter performance. Therefore, for $C1$ and $C3$ we could use 330-pF capacitors. A 680-pF unit would be suitable at $C2$.

A typical attenuation curve for a half-wave low-pass filter is given in Fig. 4.33(b). This is the measured response for the filter in Fig. 4.33(a). The second harmonic will be reduced 31 dB and the third har-

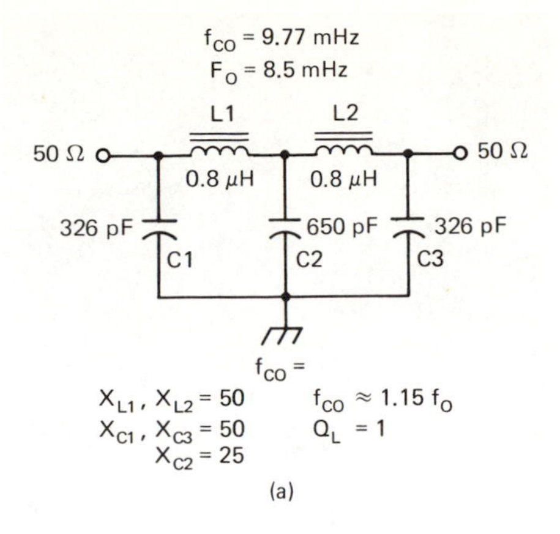

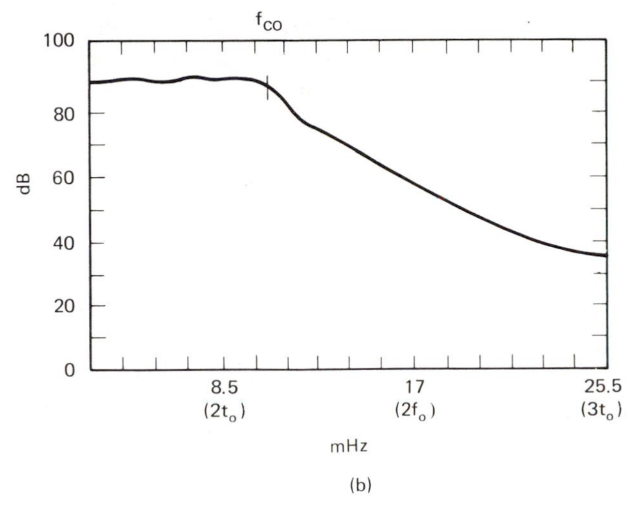

Figure 4.33 (a) Design information for a half-wave harmonic filter based on a loaded *Q* of 1; (b) typical response curve for this style of filter.

monic will be down 53 dB. Most solid-state amplifiers have heavy harmonic currents in the collector circuit, with the second and third orders seldom being less than 15 dB below peak power. Assuming the second and third harmonics from an unfiltered amplifier were 12 and 15 dB below peak power, respectively, the addition of the half-wave filter of Fig. 4.33(a) would yield a resultant second harmonic level of -43 dB and a third harmonic that is 68 dB below peak power. These values would be acceptable for most applications, although much greater suppression is the objective of careful designers. Figure 4.34 shows the spectrograph obtained from a poorly filtered transmitter. The second harmonic is down only 30 dB, and the third harmonic is even greater in amplitude, -28 dB as referenced to peak power. Furthermore, a spuri-

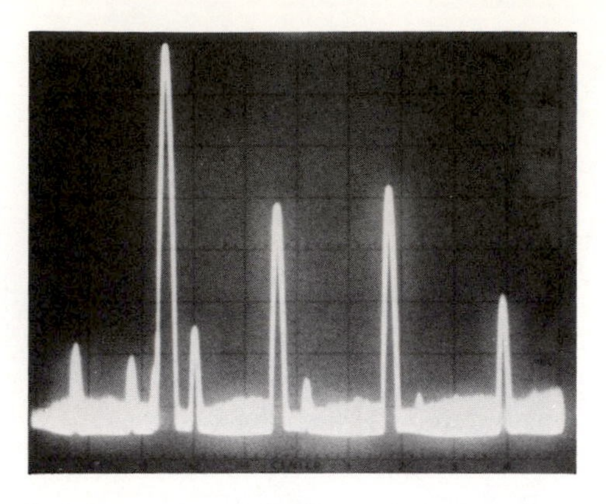

Figure 4.34 Spectrograph showing insufficient harmonic attenuation at the output of a solid-state transmitter.

ous response is evident just to the right of the carrier. Harmonics from it can also be seen near the analyzer base line. The spurious response in this example was caused by self-oscillation near the carrier frequency.

Table 4.1 contains data for fabricating Chebyshev-response low-pass filters that offer excellent attenuation of harmonic energy. Details are provided for 3-, 5-, 7- and 9-pole low-pass filters, normalized to 1 MHz and 50 Ω. The values of L are in microhenries, and the values of C are in picofarads. A specific design can be achieved by dividing the listed component values by the desired frequency in megahertz. The 1-MHz value represents the filter cutoff frequency. If impedance levels other than 50 Ω are desired, the inductance values can be multiplied by the ratio of $50/Z_o$ and the capacitor values by $50/Z_o$, where Z_o is the new impedance. An attenuation curve for the filters is seen in Fig. 4.35. It indicates the characteristics of a 5-pole filter designed from Table 4.1.

If we were to design a 5-pole filter with a ripple of 0.001 dB for the 8.5-MHz application of Fig. 4.33, the computer-derived values of Table 4.1 would call for capacitance values of 177 pF at $C1$ and $C3$, 428 pF at $C2$, and 0.9 μH at $L1$ and $L2$. Examination of Fig. 4.35 reveals that the lower the ripple factor, the poorer is the attenuation of the filter. Therefore, in the interest of good harmonic suppression, we may opt for a 1-dB ripple factor when designing our filters. A 1-dB ripple will be acceptable for the majority of our transmitter-filtering applications. Additional filter sections will sharpen the roll-off characteristic of the low-pass filter.

Toroidal inductors are recommended for the filters. For work in the medium- and high-frequency spectrums, a Micrometals type-6 powdered-iron toroid will provide good values of Q. The core size will be based on the operational flux density discussed elsewhere in this chapter.

Table 4.1

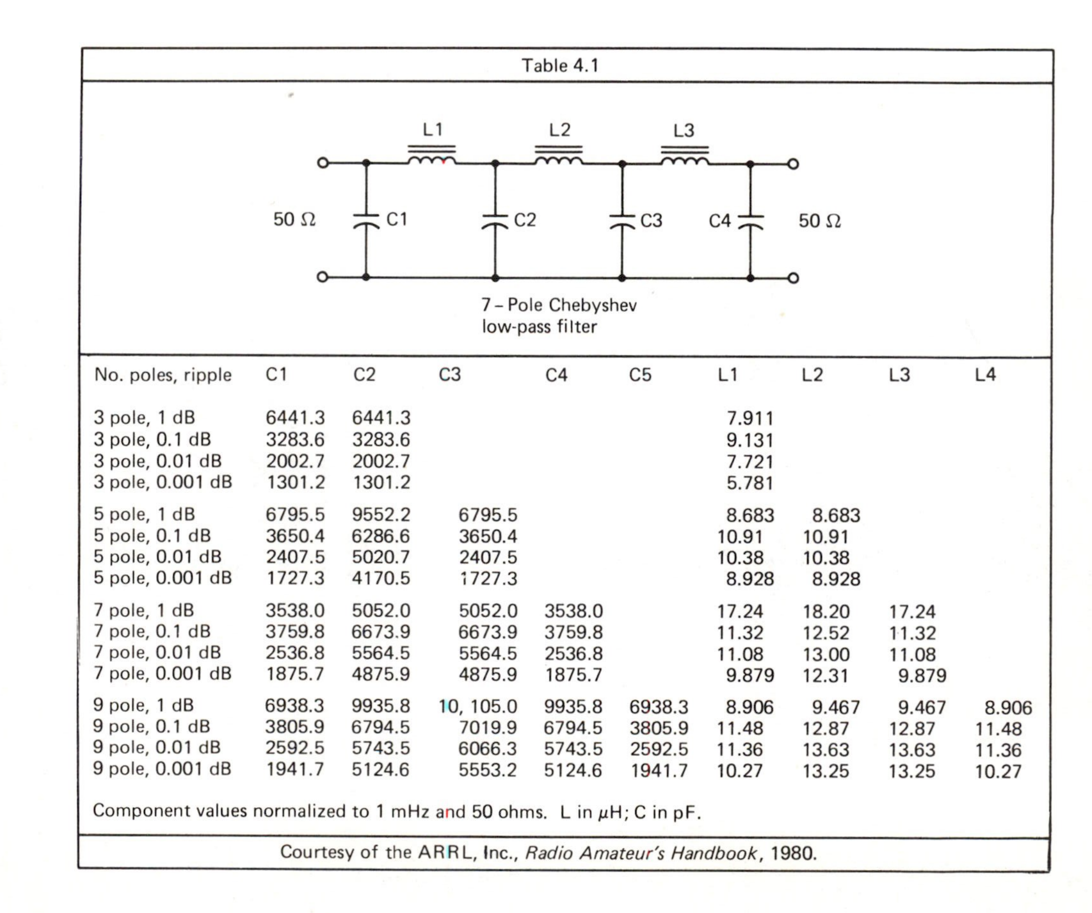

7 - Pole Chebyshev
low-pass filter

No. poles, ripple	C1	C2	C3	C4	C5	L1	L2	L3	L4
3 pole, 1 dB	6441.3	6441.3				7.911			
3 pole, 0.1 dB	3283.6	3283.6				9.131			
3 pole, 0.01 dB	2002.7	2002.7				7.721			
3 pole, 0.001 dB	1301.2	1301.2				5.781			
5 pole, 1 dB	6795.5	9552.2	6795.5			8.683	8.683		
5 pole, 0.1 dB	3650.4	6286.6	3650.4			10.91	10.91		
5 pole, 0.01 dB	2407.5	5020.7	2407.5			10.38	10.38		
5 pole, 0.001 dB	1727.3	4170.5	1727.3			8.928	8.928		
7 pole, 1 dB	3538.0	5052.0	5052.0	3538.0		17.24	18.20	17.24	
7 pole, 0.1 dB	3759.8	6673.9	6673.9	3759.8		11.32	12.52	11.32	
7 pole, 0.01 dB	2536.8	5564.5	5564.5	2536.8		11.08	13.00	11.08	
7 pole, 0.001 dB	1875.7	4875.9	4875.9	1875.7		9.879	12.31	9.879	
9 pole, 1 dB	6938.3	9935.8	10, 105.0	9935.8	6938.3	8.906	9.467	9.467	8.906
9 pole, 0.1 dB	3805.9	6794.5	7019.9	6794.5	3805.9	11.48	12.87	12.87	11.48
9 pole, 0.01 dB	2592.5	5743.5	6066.3	5743.5	2592.5	11.36	13.63	13.63	11.36
9 pole, 0.001 dB	1941.7	5124.6	5553.2	5124.6	1941.7	10.27	13.25	13.25	10.27

Component values normalized to 1 mHz and 50 ohms. L in μH; C in pF.

Courtesy of the ARRL, Inc., *Radio Amateur's Handbook*, 1980.

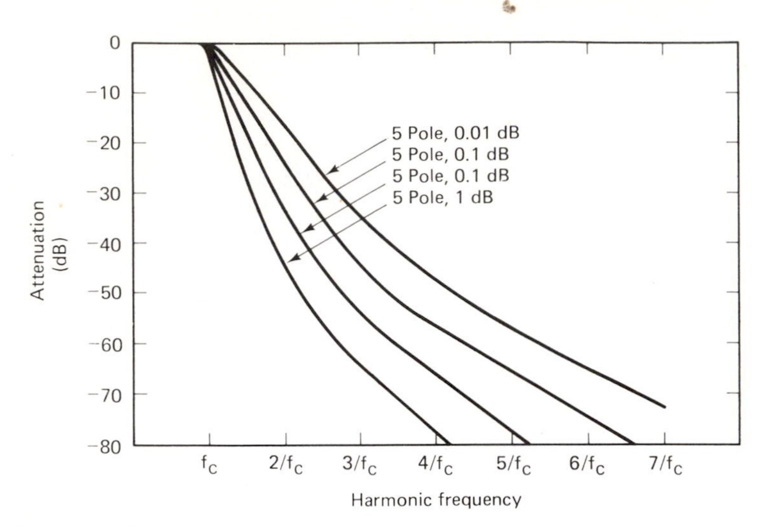

Figure 4.35 Curves showing the response characteristics of a 5-pole Chebyshev low-pass filter from Table 4.1. (Courtesy of The ARRL, Inc., *Radio Amateur's Handbook*, 1980.)

4.10.2 VHF and UHF Filtering

Lumped-constant filters of the type shown in Fig. 4.33 are difficult to design for VHF and higher frequencies. One problem is poor ultimate attenuation and another is the existence of stray inductive and capacitive reactances, which seriously complicate the filter design. A more effective solution to harmonic filtering at these higher frequencies can be found in the use of strip-line, helical-resonator, or resonant-cavity bandpass filters. Such filters yield very high values of Q, and hence their desirability at the output of a VHF or UHF transmitter (or local-oscillator chain). Strip-line filters are easy to build and are relatively inexpensive. They are effectively low-impedance sections of air-dielectric transmission line. Double-sided printed-circuit board can be used to fabricate low-power versions of these filters. Larger conductors are necessary at high power levels to reduce heating caused by circulating currents. The greater surface areas reduce RF losses in the filters. Silver-plated copper or brass (gold is even better) conductors are normally used in the construction of high-Q cavity or strip-line bandpass filters. Furthermore, the joints should be silver-soldered to enhance the Q. Surfaces joined by means of screws will cause inferior performance. An ideal filter of this variety would be entirely seamless and without electrical joints.

The mechanical and electrical aspects of quarter-wavelength strip-line bandpass filters are depicted in Fig. 4.36. The Q cannot be speci-

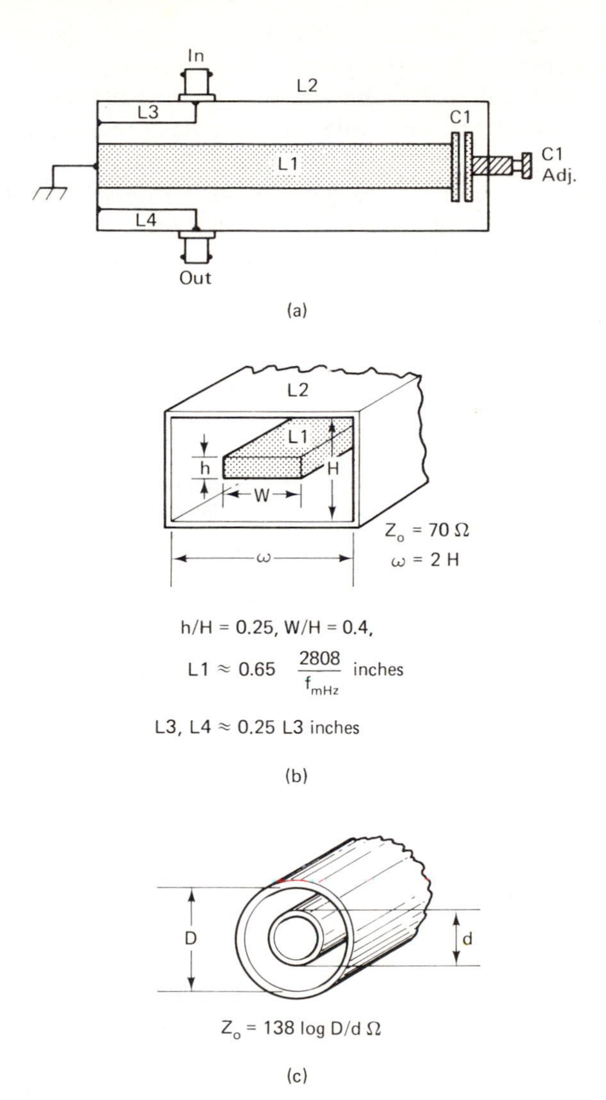

Figure 4.36 (a) and (b) Quarter-wavelength strip-line bandpass filter details for use at VHF; (c) cylindrical line sections can be used in accordance with the details given.

fied in advance, since it will depend largely on the quality of construction, the degree of load coupling, and the materials used. For the circuit of Fig. 4.36(a) the insertion loss in decibels will be governed by the placement of $L3$ and $L4$ with respect to $L1$. As the coupling links are moved closer to $L1$, the resonator Q will decrease and so will the insertion loss, giving rise to greater bandwidth. For most filtering applications the links should be set for an insertion loss of 0.1 to 0.3 dB. $C1$ is shown as a disk tuning capacitor. This style of capacitor will provide better Q than can be obtained through the use of piston capacitors or those with rotor and stator plates. It is imperative when using disk tuning that the collet for the movable-disk shaft be electrically and mechanically positive. This calls for quality machining, fine threads, and some type of tensioning of the movable half of the capacitor. The $C1$ shaft will require a locking mechanism to keep the resonator tuned to the correct frequency, since the very high Q of the circuit can lead to detuning when the assembly is subjected to vibration or shock. $L1$ is made shorter than a quarter-wavelength to provide sufficient tuning range via $C1$. Highest Qs will generally be realized with the greatest practical L-to-C ratio.

An arbitrary 70-Ω line characteristic for $L1$ and $L2$ of Fig. 4.36(a) is specified for the design. Figure 4.36(b) includes details for determining the line dimensions in inches (mm = in. × 25.4). If the rectangular format is not used, the line can be tailored cylindrically. Details are given in Fig. 4.36(c).

For operation above 200 MHz it is more convenient to adopt the half-wavelength strip-line configuration. This is seen in Fig. 4.37. At

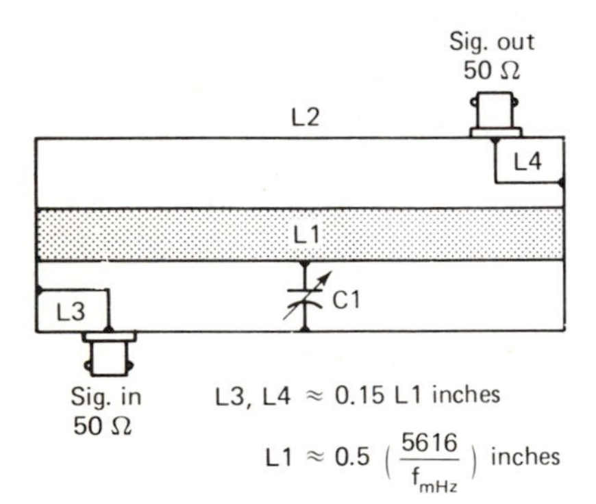

Figure 4.37 Half-wavelength strip-line bandpass filters are suitable for use at UHF.

UHF it is not uncommon to encounter cavity resonance in addition to $L1/C1$ resonance when using the quarter-wavelength format. This results in a double response that can ruin the filter performance. This effect becomes particularly troublesome in the 450-MHz region if extreme care is not taken. The half-wavelength design permits relative freedom from the effects of $L2$ cavity resonance near the frequency of operation. The h, H, W, and W dimensions for the filter of Fig. 4.37 are the same as for the quarter-wave filter of Fig. 4.36(a) and (b). $C1$ is at the electrical center of $L1$.

Lumped LC bandpass filters are effective below approximately 200 MHz when constructed as shown in Fig. 4.38. Coupling to the load is effected by selecting the appropriate tap points on coils $L1$ and $L2$. Coupling between the resonators is accomplished by virtue of the wall aperture labeled "A." As the aperture size is decreased, the Q will rise and the insertion loss will increase. A loss of 0.1 to 0.3 dB is satisfactory for most filters of this type. The trade-off must always be between

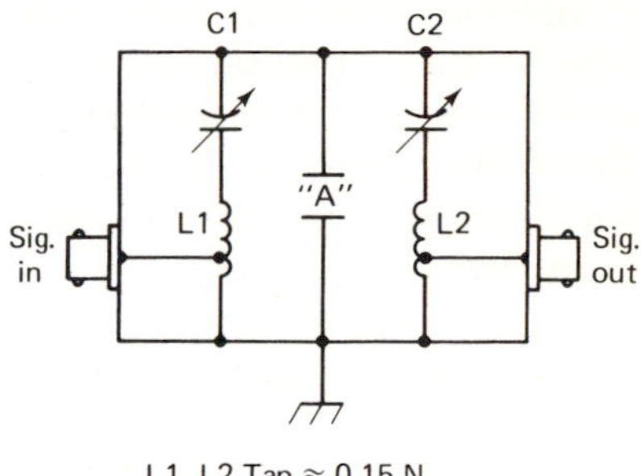

Figure 4.38 Lumped-constant bandpass filters are effective for harmonic filtering at VHF.

gain and selectivity, since no passive filter is without losses if, indeed, it is a filter.

The highest *Q*s for the circuit of Fig. 4.38 will be obtained when *L*1 and *L*2 are made from large-diameter wire or tubing that has been silver plated. Similarly, the enclosure and aperture partition should be silver plated and the joints formed by means of silver solder. *L*1 and *L*2 should be spaced approximately one coil diameter from the box walls in the interest of high *Q*. *C*1 and *C*2 need to be of high dielectric quality for best resonator *Q*. Although this circuit closely resembles the *helical resonator*, it does not yield the extremely high *Q* available from that kind of filter. Design tables and equations for helical resonators are given in *Reference Data for Radio Engineers* (Howard W. Sams & Co., Inc., Indianapolis, 1975), and in the *ARRL Electronics Data Book* (ARRL, Inc., Newington, Connecticut, 1976).

4.11 PRACTICAL AMPLIFIERS

It would be impossible for us to follow a recipe-book design procedure in RF power-amplifier work. The fundamentals of design have been offered in the foregoing sections of this chapter. Each amplifier will function differently during initial tests, even though we adopt a given design and make several amplifiers from it. This is because there will be differences in each transistor from a given production run, and because component values and lead lengths will not be precisely the same in each otherwise identical amplifier. Therefore, it is necessary that we optimize the amplifier after it is assembled. The notable exception to this rule is when circuit boards are mass produced to close tolerances, the transistors are graded and matched, and the small components are of close tolerance. Even under these conditions we may find it necessary to do some final refining of the performance.

In this section are some practical circuits that were built and used by the author. They serve mainly as general examples that illustrate the concepts we have discussed in this chapter. The bibliography at the end of this section provides a wealth of additional reference material for the reader, and specific design data are offered for operating frequencies not covered here.

4.11.1 Class C VMOS Amplifier

Figure 4.39 contains the circuit of a class C RF amplifier that delivers up to 20 W of output with an efficiency of 80%. Only 100 mW of

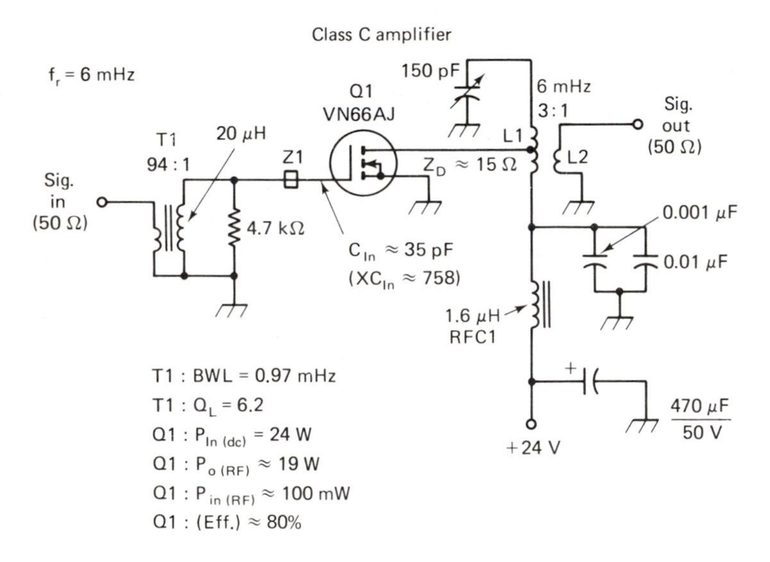

Figure 4.39 Practical circuit for a class C VMOS FET amplifier.

driving power is needed to develop the required V_{gs} to fully excite the amplifier.

The input circuit has been broadbanded by using *T*1 in combination with the 4.7-kΩ shunt resistor. *T*1 is broadly resonant at 6 MHz by virtue of C_{in} of *Q*1 (35 pF). The input circuit has a bandwidth of approximately 1 MHz because the resultant Q_L is 6.2. *Z*1 has been used to discourage VHF parasitic oscillations. *T*1 has a turns ratio of 10 : 1, since a 9.7 : 1 ratio would be impractical to realize.

The amplifier drain is tapped down on *L*1 to improve the resonator *Q* and to effect a proper impedance match to the load. The *Q*1 drain characteristic will be roughly 15 Ω, based on $Z = V_{DD}^2/2P_o$. The turns ratio between the bottom section of $L1$ and $L2$ will be 1.7 : 1 by calculation, but a 2 : 1 ratio will be satisfactory. $RFC1$ and the associated bypass capacitors function as a decoupling network in the V_{DD} supply line. Linear operation of this amplifier can be achieved by lifting the low end of the $T1$ secondary winding and applying forward bias to the transistor gate. A level of +2 to +4 V will be required. This design can be used as a model for VMOS FET amplifiers that deliver greater P_o and use larger transistors. A harmonic filter is necessary at the output of this amplifier to ensure spectral purity. Linear operation will result in harmonic currents of less magnitude than will be the case during class C operation.

4.11.2 Single-Ended VHF Amplifier

The Motorola MRF200-series bipolar transistors are capable of delivering smooth performance at VHF and UHF. A practical design that operates in the 200-MHz region and uses a MRF209 transistor is shown schematically in Fig. 4.40. An output of 25 W is available with 7 W of drive. Amplifier gain is approximately 6 dB with this circuit. $C1$ and $C2$ are 100-pF ceramic chip capacitors that were added to aid stability and reduce harmonic currents.

Class C operation will occur when the bottom end of $RFC1$ is grounded. Linear amplification can be had by applying forward bias to $Q1$ through $RFC1$. This circuit is intended mainly for amplification of FM signals, which does not require linear class A or class AB operation

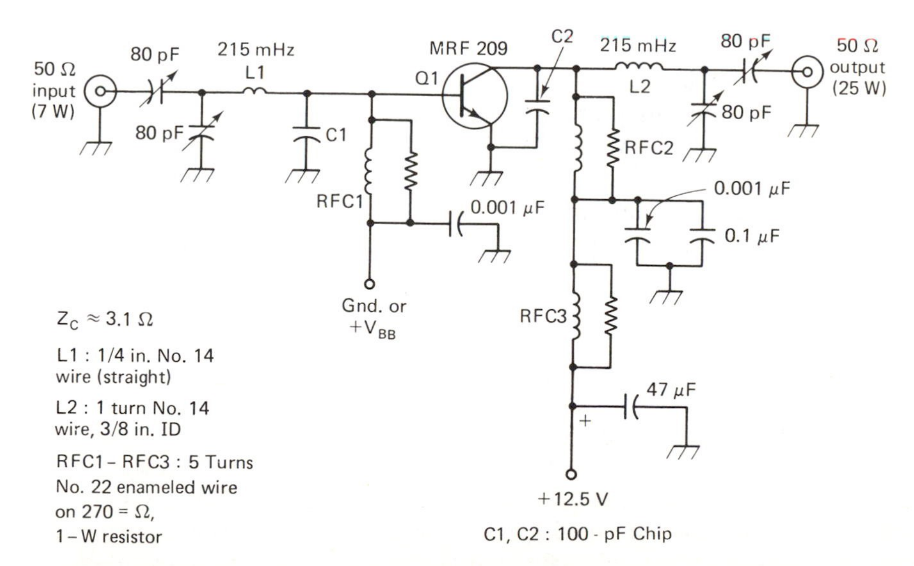

Figure 4.40 Circuit of a single-ended bipolar-transistor amplifier that delivers 25 W of power at 215 MHz.

of $Q1$. The emitter connection to ground needs to be very short and direct to discourage lead inductance, which can cause unwanted degenerative feedback. Assembly on a double-sided pc board [see Fig. 4.29(b)] will aid stability and overall performance. Strip-line inductors can be substituted at $L1$ and $L2$ if high-quality glass-epoxy circuit-board material is used.

$RFC1$ to $RFC3$, inclusive, are wound on the bodies of 270-Ω, 1-W carbon resistors. The coils are connected electrically to the resistor pigtails to form a parallel RL circuit. This will lower the choke Q and aid stability at lower frequencies. Lower values of resistance can be used if instability is observed. Values as low as 18 Ω will not impair performance at 215 MHz.

4.11.3 One-Hundred-Watt VMOS Amplifier

Broadband linear amplification is especially simple when the active devices are VMOS power FETs. The feedback network becomes fairly complicated when designing a similar amplifier around bipolar transistors. This is because the bipolar devices exhibit large changes in terminal impedance with frequency, as discussed earlier in the chapter. The input capacitance of bipolar transistors increases markedly as the operating frequency is lowered. This undesirable condition does not prevail when using VMOS devices. Thus, we can design our feedback network and have it function the same over the entire amplifier bandwidth. Additionally, the VMOS amplifier will not be subject to second breakdown and thermal runaway, another advantage over the bipolar-transistor amplifier.

Figure 4.41 shows how a 100-W output broadband VMOS amplifier can be configured for linear service. Removal of V_{GS} bias and grounding that circuit point will place the amplifier in the class C mode for CW, RTTY, and FM use. An input VSWR of less than 2 : 1 across the 30- to 88-MHz range is obtained with the transformer arrangement shown. Transmission-line transformers are used for $T1$ to $T6$, inclusive. They can be wound on ferrite balun or toroid cores. $C1$ is a reactance-compensating trimmer. It should be adjusted for the lowest input VSWR.

For linear operation the V_{GS} bias should be set for a quiescent drain current ($Q1$, $Q2$ total) of 2.2 A. This should occur in the 2- to 4-V range at the gates of the transistors.

Shunt feedback is obtained by bridging the 1000-Ω, 2-W resistors from drain to gate. $R1$ and $R2$ establish the input load seen by $T1$, $T2$, and $T3$ from the signal source and form part of the feedback network. $R1$ and $R2$ can be fashioned from three 18-Ω, 2-W carbon resistors in a parallel hookup. This will help reduce the stray inductance while yield-

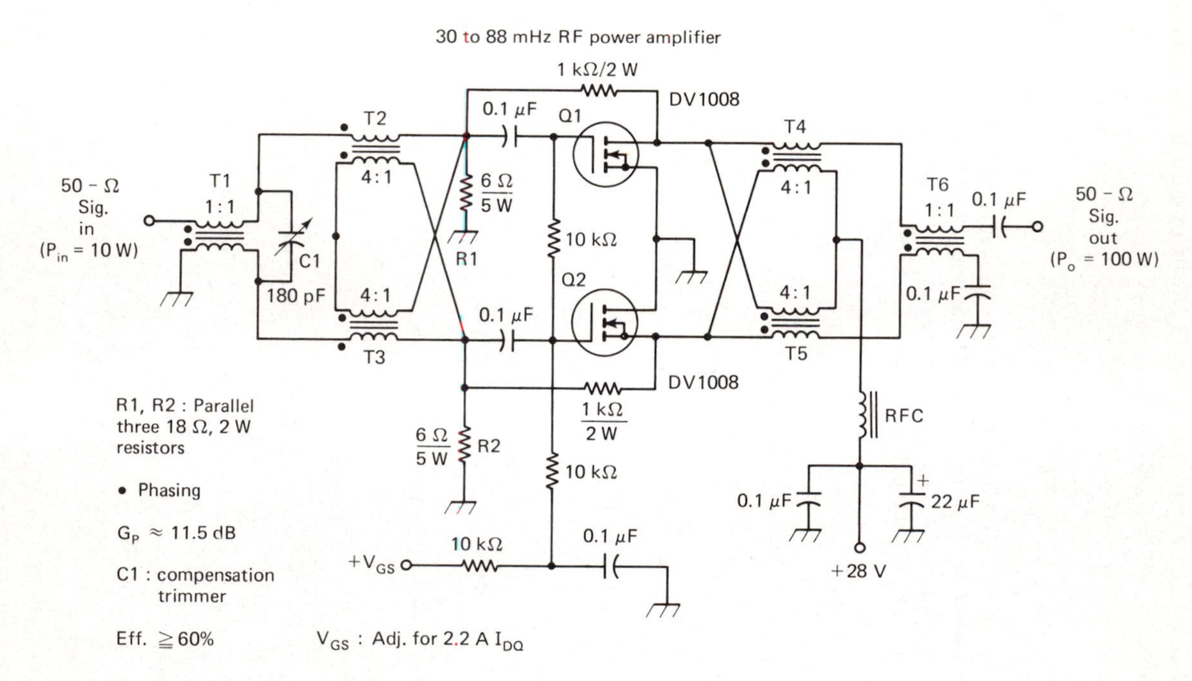

Figure 4.41 A 100-W push-pull VMOS FET amplifier for broadband linear applications. (Courtesy of Siliconix, Inc.)

ing the desired 6-Ω characteristic at each gate. Since they will always be slightly reactive (X_L), they complement the feedback circuit.

Effective harmonic filtering must follow this amplifier in the interest of acceptable spectral purity. A 5-pole Chebyshev filter, based on the data in Table 4.1, is recommended. Amplifier gain is approximately 10 dB, inclusive of transformer losses.

BIBLIOGRAPHY

Additional references are given here to provide the reader with expanded coverage of the subjects treated in Chapter 4. Most of the books listed are standard parts of the engineer's or technician's library and are highly recommended to the reader in the event he or she does not already have access to them.

Application Notes

Becciolini, "Impedance Matching Networks Applied to RF Power Transistors," Motorola *AN-721.*

Davis, "Matching Network Design with Computer Solutions," Motorola *AN-267.*

Granberg, H., "Measuring the Intermodulation Distortion of Linear Amplifiers," Motorola *EB-38,*

____, "A Two-Stage 1-KW Solid-State Linear Amplifier," Motorola *AN-758.*

____, "Broadband Transformers and Power Combining Techniques for RF," Motorola *AN-749.*

Hejhall, R., "Understanding Transistor Response Parameters," Motorola *AN-139A.*

Leighton, L., and E. Oxner, "AGC for the VMOS RF Power Amplifier," Siliconix *AN80-6.*

____, "VHF Power Amplifier Design Using VMOS Power FETs," Siliconix *AN80-4.*

Books

ARRL, Inc., *The Radio Amateur's Handbook*, transistor theory and practical circuits, Newington, Conn., 1980, 1981.

DeMaw, D., *Practical RF Communications Data for Engineers and Technicians*, No. 21557, Howard W. Sams, Inc., Indianapolis, 1978.

____, *Ferromagnetic-Core Design and Theory*, RF Applications, 1981, Prentice-Hall, Inc., Englewood Cliffs, N.J., 1981.

Hayward, W. and D. DeMaw, *Solid State Design*, 1977, ARRL, Inc., Newington, Conn., 1977.

Johnson, J., *Solid Circuits*, No. 2.2.8.OA, Communications Transistor Corp. (CTC),

Motorola, Inc., *Motorola RF Data Manual*, 1978. Phoenix, AZ.

RCA, *Solid-State Power Circuits,* No. SP52, Somerville, N.J.

Papers

DeMaw, D., "An Experimental VMOS Transmitter," *QST Magazine*, May 1979.

Hayward, W., "A VMOS Transmitter for 10-Meter CW," *QST Magazine*, May 1979.

Leighton, L., "HF Power Amplifier Design Using VMOS Power FETs," *RF Design*, January 1980.

Oxner, E., "Power FETs," *Wireless World*, May 1977.

5

FREQUENCY MULTIPLIERS

Frequency multiplication becomes necessary in a host of communications circuits for the purposes of overall circuit efficiency and reaching an operating frequency that cannot be realized by VFOs and crystal oscillators. With the latter, a limitation is imposed by the state of the crystal art: the upper limit for practical quartz crystals is on the order of 20 MHz. Therefore, if we are to have the advantage of good frequency stability at frequencies above the practical level of crystal oscillators, frequency multipliers become a useful tool in our design work.

We can increase the fundamental frequency by means of passive (diode) or active (transistor) multipliers. The essential difference between the two kinds of multipliers is that a passive multiplier will provide a smaller output power than the input amount, whereas an active multiplier can produce output powers greater than the input power, depending on the operating parameters.

Although single-stage, or single-ended, active multipliers are commonly used in frequency-multiplication applications, they are not especially efficient. Generally, a single-ended doubler has an efficiency of roughly 50%, a tripler yields about 30%, and a quadrupler approximately 25%. If we keep these numbers in mind, we can plan our gain distribution accordingly, thereby obtaining the required power output at the final operating frequency.

Vacuum-tube frequency multipliers rely upon envelope distortion to obtain the necessary harmonic currents for frequency multiplication. The desired end result is enhanced considerably by using transistors in multipliers. This is because the harmonic energy is partly the result of envelope distortion and the nonlinear action of the transistor junction during the wave period. Since nonlinearity is conducive to the generation of harmonic currents, the semiconductor offers a distinct advantage in frequency multiplication. Furthermore, a frequency multiplier, from a truly classical viewpoint, should operate in class C to obtain the desired result. Therefore, we may tend to rule out the use of class A or B modes for harmonic generation. Truly, the class C mode is the preferred one for our purpose, but only if we have sufficient driving power to justify class C operation. A bipolar transistor can be biased for class A service and still deliver a substantial amount of harmonic power by virtue of the nonlinear action of the junction capacitance change during the excitation period. The efficiency will not be as high as that of a similar stage operated in class C, but for many applications it will prove adequate.

There is still a third condition to consider: a transistor can be biased to operate class A as a multiplier, but can be driven well into the class C region. The effect of the forward bias is to make the driving-power requirement lower than for a stage that has no forward bias applied. The choice of our operating mode will sometimes be founded on available excitation power and the required power output from the multiplier stage.

5.1. FREQUENCY DOUBLERS

Perhaps the most common type of frequency multiplier we will be working with when designing transmitters or local-oscillator chains in receivers and converters is the doubler. It can take many forms, and the choice is anything but arbitrary. The full-wave doubler offers the best performance in terms of efficiency and suppression of the driving frequency. Our emphasis will be placed on that type of circuit in this section.

One of the more simple circuit forms for frequency doubling is offered in Fig. 5.1. The small-signal dual-gate MOSFET shown is capable of functioning rather well as a push-push type of doubler by applying push-pull excitation to gates 1 and 2, as shown. In effect, we are using *Q*1 as a mixer by injecting the gates with 15-MHz energy, 180° out of phase, respectively. The net effect is the application of two 15-MHz signals at the stage input, with the sum frequency (30 MHz) appearing at the output. For many low-level doubling applications, we may find this circuit entirely suitable. Suppression of the 15-MHz energy will be excellent (−40 dB or greater) if care is taken to keep the input drive symmetrical, and if the drain tuned circuit has a reasonably high loaded *Q*. Inclusion of a source-bias resistor and bypass capacitor

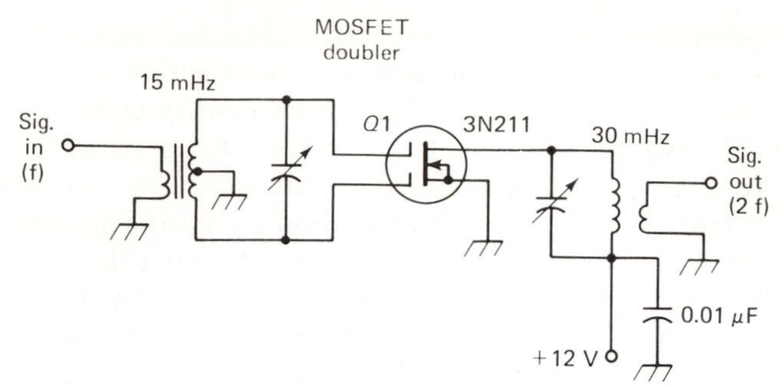

Figure 5.1 Circuit of a simple frequency doubler that uses a small-signal dual-gate MOSFET.

may in some instances (depending on the MOSFET characteristics) improve the doubler efficiency.

Discrete transistors (bipolars or FETs) can be used successfully in a balanced push-push doubler as illustrated schematically in Fig. 5.2. Ideally, $Q1$ and $Q2$ would be matched closely for dynamic characteristics. Since this is seldom practical, we can effect reasonably good balance, and hence good suppression of the driving frequency, by including $R1$ in the circuit. It can be adjusted, with excitation applied at the doubler input, for minimum driving frequency (f) at the output of $T2$. Compensation for imbalance at the doubler input can be accommodated through the inclusion of $C1$. This trimmer capacitor may be required at the gate of $Q1$ or $Q2$, depending on the layout and stray capacitances that will be present. $C1$ is added at the transistor gate where it has the desired effect, and this is done experimentally.

A broadband, trifilar-wound input transformer ($T1$) ensures that

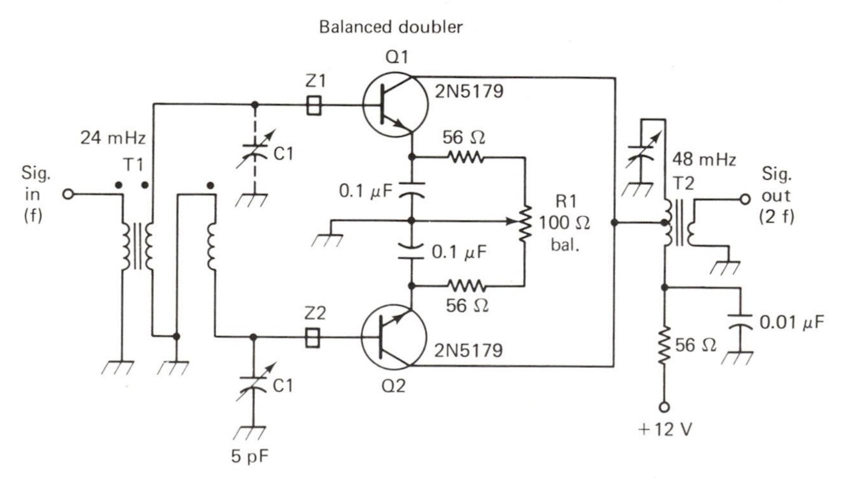

Figure 5.2 Discrete transistors can be used effectively as push-push doublers by obtaining dynamic balance through use of a variable emitter resistance ($R1$).

the 24-MHz energy is fed to $Q1$ and $Q2$ of Fig. 5.2 with a phase difference of 180°. $Z1$ and $Z2$ are high-mu ferrite beads that are placed as close to the $Q1$ and $Q2$ bodies as possible. These impedances will discourage the tendency toward VHF parasitic oscillation, encouraged by the very high f_T of the 2N5179 devices. Doublers, and even mixers, will break into self-oscillation under favorable conditions, despite the input and output being on different frequencies.

$T2$ of Fig. 5.2 is a high-Q resonator. The collectors of $Q1$ and $Q2$ are tapped down on the transformer primary to minimize loading effects, thereby enhancing the loaded Q of the tuned transformer. Suppression of the driving frequency should be on the order of -50 dB if the doubler is well balanced and laid out to prevent unwanted coupling between the input and output halves of the circuit. Greater rejection of the input frequency can be achieved by following $T2$ with a bandpass filter or by making $T2$ a part of such a filter. A two-section bandpass filter can yield an output spectrum that shows all spurious responses down by -70 dB or more below peak power at the desired frequency. The two 56-Ω emitter resistors are used to prevent excessive current flow in $Q1$ or $Q2$ in the event $R1$ is set all the way at one extreme or the other of its range when operating voltage and drive are applied.

A passive full-wave doubler circuit is seen in Fig. 5.3. It can be compared to a conventional full-wave power supply rectifier, wherein the 60-Hz line frequency is doubled to 120 Hz at the rectifier output. As is true of the push-push doubler shown in Fig. 5.2, balanced drive and symmetrical layout are important if the excitation frequency is to be minimal at the doubler output. In the circuit example we see that MBD501 hot-carrier diodes are used at $D1$ and $D2$. They provide excellent switching action and are recommended for applications above approximately 30 MHz. The doubler output is filtered with a bandpass, strip-line resonator ($FL1$) to suppress the driving frequency and harmonics of $2f$. Owing to the power loss through $D1$ and $D2$, plus the insertion loss of the bandpass filter, a post-doubler amplifier ($Q2$) is

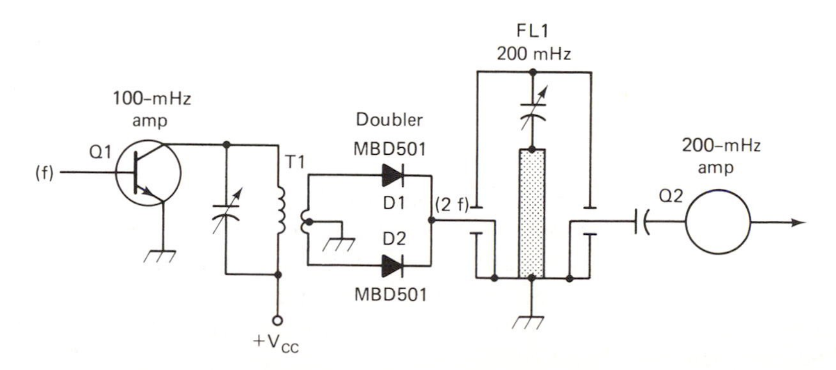

Figure 5.3 Full-wave frequency doubler using diodes and a bandpass filter.

added to develop the required signal level for the succeeding stage. Suppression of the driving frequency and high-order harmonics should be at least - 70 dB with the circuit shown, assuming that $FL1$ is adjusted for a trade-off between Q_L and power transfer. For operation below approximately 100 MHz, it will be more practical to employ a lumped LC filter at $FL1$. Also, if ample driving power is available from $Q1$ to provide good doubling action, $T1$ can be replaced by a broadband transformer such as that in Fig. 5.2 ($T1$). The subtle advantage of using a tuned transformer ahead of the diode doubler is that harmonics of the input frequency, generated by $Q1$, are not passed along to the doubler in sufficient magnitude to cause additional responses from the doubler.

We can call upon one of the transistor-array ICs for use as push-push doublers if we desire good dynamic balance. Such an array is illustrated in Fig. 5.4. Figure 5.4(a) shows the inner workings of the RCA CA3724G chip, which contains four *npn*, high-frequency silicon transistors. Good balance is ensured by virtue of the transistors being formed at the same time on a common substrate. Also, they will maintain their electrical balance with changes in temperature, since each device will track as temperature excursions occur.

In Fig. 5.4(b) we have cascaded two push-push doublers to provide an output frequency of $4f$ from the IC. The emitters are grounded, and no forward bias is applied to the transistors. This results in class C operation for maximum efficiency. Enhancement of the circuit balance may be effected by adding a 5-pF trimmer at pins 2 or 13 and pins 6 or 9 of $U1$. The most effective placement of the trimmer will be determined in the same manner as when working with the circuit in Fig. 5.2.

There are a number of RF types of ICs that can be used as push-push doublers. The IC used in Fig. 5.4 is capable of delivering a fair amount of output power because it will dissipate safely a current maximum of 1 A. Each transistor is rated at 1 W maximum at a case temperature up to 25°C. The maximum dissipation for the entire package is 2 W. The V_{ce} maximum is 40 V. Turn-on time is 30 ns at 500 mA, and turn-off time is 36 ns at 500 mA, I_c.

A lower-power IC that will perform very well as a push-push doubler is the RCA CA3028A. It contains a differential transistor pair and a bipolar-transistor current source. It was designed primarily as an IF amplifier in the high-frequency range, but works nicely as a doubler up to approximately 100 MHz. To program the CA3028A for class C operation, it is necessary to ground the emitter of the current source and apply sufficient forward bias to the current source to saturate it. This removes the current source from the circuit for practical considerations and places the emitters of the differential pair effectively at ground. Push-pull drive is applied to the bases of the differential pair, and $2f$ is taken from the paralleled collectors.

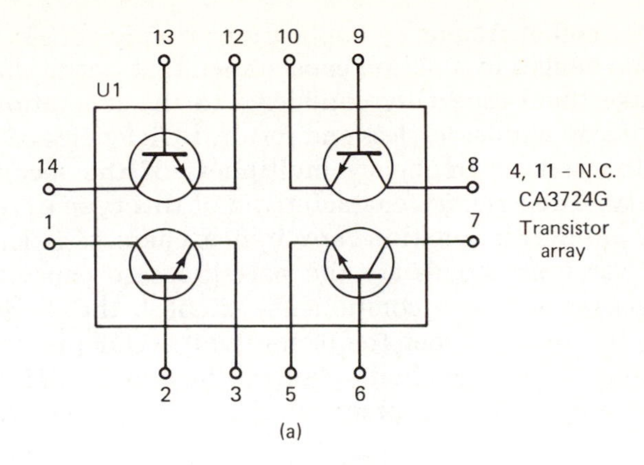

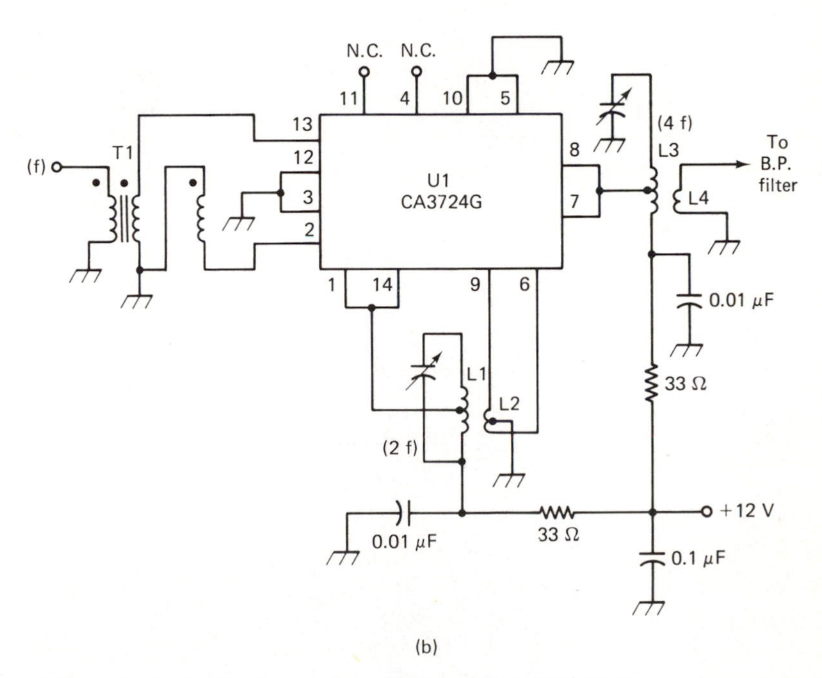

Figure 5.4 (a) Inner workings of an RCA CA3724G transistor array IC; (b) cascaded push-push doublers are used to obtain an output of $4f$ by using the CA3724G IC.

5.1.1 Varactor Multipliers

Frequency multiplication at moderate power levels is achieved effectively by means of varactor diodes of the abrupt-junction type. In effect, the varactor used by modern-day designers is referred to as the "charge storage" or "step-recovery" diode. The *pn* junction is made of silicon, or for higher-frequency applications it may consist of gallium arsenide. The changes in junction capacitance that occur with reverse bias applied make them especially conducive to the generation of harmonic currents with an almost lossless end result. Efficiencies of 70% are quite typical with varactor frequency multipliers of the second- and third-order variety. The primary characteristic of this type of diode is high Q and a very nonlinear junction-capacitance characteristic. The limiting factors of varactor diodes are the parasitic case capacitance and the junction resistance. These components establish the diode transit time, and hence the useful upper frequency limits. Our present state of the art has produced varactors that operate as high as 20 GHz.

Figure 5.5 contains a practical circuit for a varactor tripler that

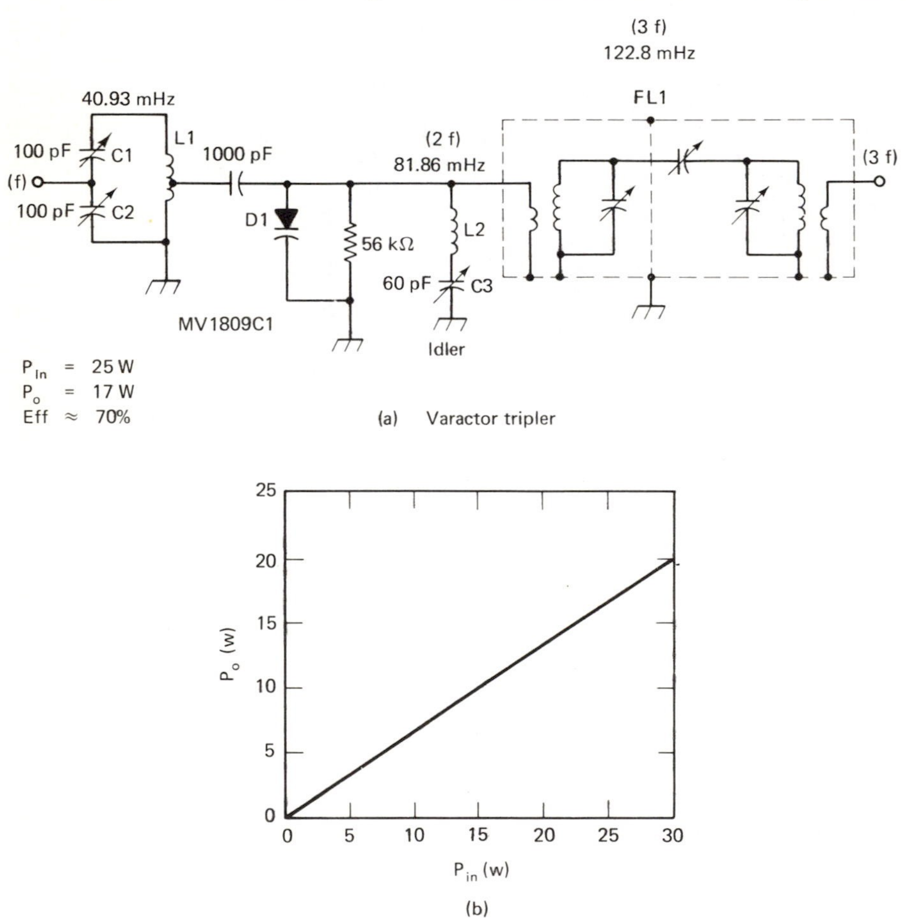

Figure 5.5 (a) Varactor tripler with output on 122.8 MHz; (b) characteristic curve showing power input versus power output for a typical Varactor multiplier.

is capable of delivering 17 W of output power at 122.8 MHz. The MV1809C1 diode specified at $D1$ will provide equivalent power output up to 2 GHz. $C1$ and $C2$ are adjusted for an input VSWR of 1 with drive power applied, and with a load attached to the output of $FL1$.

$L2$ and $C3$ comprise the "idler tank," which is tuned to $2f$. This circuit enhances the generation of $3f$ by virtue of mixing the f and $2f$ currents to obtain the sum frequency of $3f$. This, plus the nonlinear change in junction capacitance, creates high-level $3f$ currents at the input to $FL1$. Filtering by means of $FL1$ is essential in the interest of output purity. A well-designed cavity or strip-line bandpass filter is often used in place of the lumped LC filter shown in Fig. 5.5.

With the circuit shown, an unfiltered output of 17 W can be obtained with 25 W of input power at $C1/C2$. The actual output from the circuit will be determined by the insertion loss of $FL1$, which to function as a filter must have some loss. An Amperex H4A varactor can be used in place of the MV1809C1 diode for operation in the VHF and UHF regions. The MV1809C1 is intended primarily for use in the microwave region.

Figure 5.5(b) shows a representative P_{in}/P_o curve for a varactor diode operating in the power range under discussion. Although the curve relates in this example to a varactor tripler, the power-input/power-output characteristic will be essentially the same for a varactor doubler. Varactors have been used successfully up to the eighth harmonic of the driving frequency, although as the multiplication order is increased beyond the third, efficiency declines progressively. Nonetheless, the technique permits the inexpensive generation of frequencies at high-order harmonics into the microwave spectrum.

For low-power applications it is practical to place several small-signal, high-speed switching diodes in parallel (such as 1N914s) to achieve varactor action. Good results can be obtained with this procedure through at least 1.5 GHz if the lead lengths are kept short and the stray capacitance is minimized. The transit time (assuming parasitic capacitance is kept low) will be reasonably good as a result of paralleling several diodes to reduce the forward resistance of the composite varactor.

The primary drawback in using varactor frequency multipliers is the *frequency sensitivity* of that style of circuit. We will learn in adjusting a varactor multiplier that the tuning is rather critical with respect to interaction between the various adjustments. Optimum output power will occur over a narrow band of frequencies. Significant changes in frequency (determined in part by the network Qs) will require retuning of the multiplier if maximum efficiency is to result. Also, power varactors dissipate a substantial amount of heat, which requires effective heat sinking, just as we find necessary when using power transistors.

5.2 HETERODYNE FREQUENCY GENERATION

A viable alternative to small-signal generation of high-order frequencies is available to us through a process of heterodyning two frequencies to produce a third (desired) one. The concept is applied regularly to transmitters and receivers that must cover a wide range of frequencies from VLF through the UHF regions.

This approach is especially important when variable-frequency control is desired, rather than channelized control. This entails the use of crystal-controlled oscillators and a variable-frequency oscillator. The two signal-source outputs are routed to a mixer, filtered, and supplied to a specified injection point as the sum or difference frequency of the two signal sources. A block diagram of the general arrangement is given in Fig. 5.6. It represents an oversimplification of a properly designed circuit.

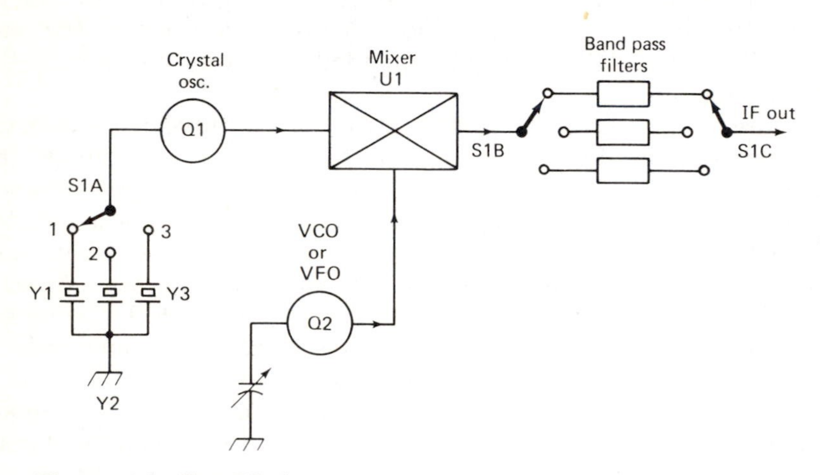

Figure 5.6 Simplified block diagram of a heterodyne-frequency generator.

A schematic diagram that shows a heterodyne generator in the most basic form is offered in Fig. 5.7. Normally, $Q1$ would be followed by one or two buffer stages to minimize frequency pulling. We might even elect to include a harmonic filter at the output of the VFO buffer stage, thereby ensuring that only the desired frequency would be injected into the mixer, $Q3$.

Depending upon the stability requirements of the equipment in which the generator would be used, a stage of buffering might be included between the heterodyne oscillator ($Q2$) and the mixer. In either case, regulated operating voltage would be a good choice at $Q1$ and $Q2$ of Fig. 5.7. This would also aid frequency stability.

The mixer shown in Fig. 5.7 is not capable of good isolation between its ports and would be used only in the most casual of designs for

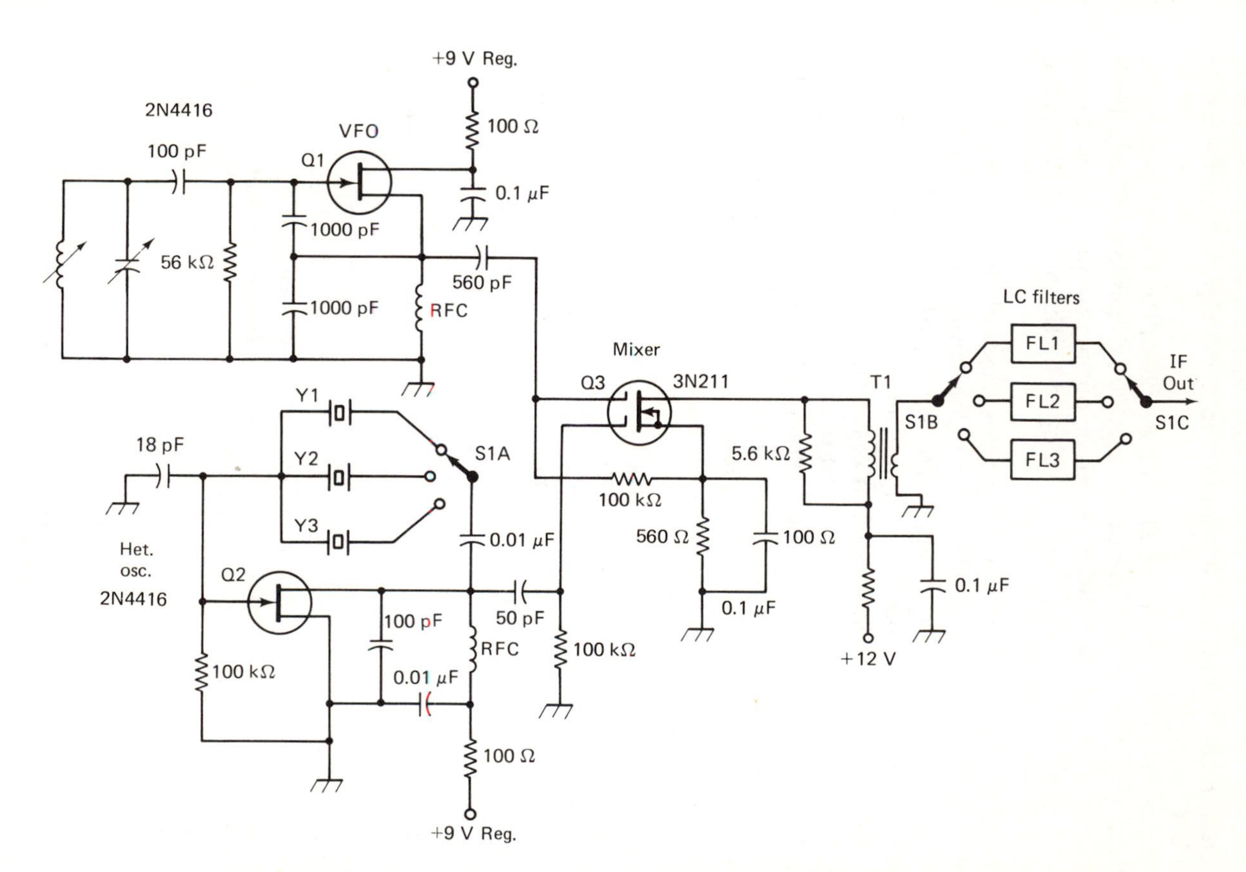

Figure 5.7 Schematic diagram of a basic heterodyne-frequency generator. A practical circuit would be somewhat more complex than shown here.

that reason. A doubly balanced active or passive mixer is the usual choice of the designer in the interest of minimum spurious responses and good dynamic range. Information on mixer design and performance is given in Chapter 6.

Output from the mixer at $f1 + f2$ or $f2 - f1$ is routed through a broadband transformer to a set of switchable *LC* bandpass filters. They will remove most of the unwanted mixer output frequencies and contribute to purity of the waveform developed by the overall heterodyne generator.

*Q*2 is shown as a Pierce oscillator for use with fundamental crystals below approximately 20 MHz. For higher orders of IF output from the mixer, we could make *Q*2 an overtone type of oscillator to permit using much higher *Q*2 injection frequencies at the mixer, *Q*3. For example, suppose we were interested in extracting a tunable IF of 48 to 50 MHz from *Q*3. Assume that *Q*2 is an overtone oscillator that permits the crystal to oscillate on its third overtone. If we design the VFO to cover a 2-MHz range, we will be able to provide an IF of 48 to 50 MHz by selecting the correct crystal frequency at *Y*1. We will use the sum frequency of the two signal sources, so we will design the VFO to tune from 5 to 7 MHz. Therefore, *Y*1 will have to provide output from *Q*2 on 43 MHz. Thus, 5 + 43 = 48 MHz, and 7 + 43 = 50 MHz.

We can take a different approach in our design to cover the same IF range. Four third-overtone crystals could be employed at *Q*2 to yield four 500-kHz switchable IF tuning ranges. If we adopted this technique, it would be necessary for the VFO to tune only 500 kHz rather than 2 MHz. It would also provide much greater bandspread on an analog frequency-readout dial. This technique is the preferred one for many commercial receivers and transmitters operating in the high-frequency spectrum.

Mechanical switching (*S*1*A*, *S*1*B* and *S*1*C*) can be replaced by diode switching to simplify the mechanical layout of the equipment. A phase-locked loop (PLL) would be a good substitute for the VFO shown in Fig. 5.7, as it would offer improved frequency stability. We can see from the foregoing that the circuit shown is indeed basic and quite flexible with regard to design philosophy. It was used merely to demonstrate the principles of heterodyne frequency generation.

6

MIXERS, BALANCED MODULATORS AND DETECTORS

There is little difference between the functions of mixers, balanced modulators, and detectors in standard communications circuits. All three operate on the principle of combining two frequencies to produce a third one. The design goals are, therefore, quite similar. We are interested in having any of the three circuits perform with high dynamic range, minimum distortion, and good suppression or isolation between the various ports of the circuit device.

We have a number of choices open to us when choosing a particular mixer, modulator, or detector, and the best circuit for the application will be founded on cost effectiveness, performance standards, and available circuit space in the composite product. In all examples we can expect optimum performance from balanced circuits as opposed to single-ended circuits. We need to make a choice between passive and active versions of the three types of circuits, since passive devices yield a conversion loss, and active circuits usually ensure a conversion gain. It should be said, however, that an active mixer, modulator, or detector can exhibit unity gain, or even a conversion loss. This will depend upon how the device is used in the circuit, as we will learn later in the chapter. Figure 6.1 provides a relative comparison between the three circuits we shall examine in this discussion.

6.1. MIXERS

The most common use of mixers is found in receivers and heterodyne-frequency generators in transmitters. We have our choice between

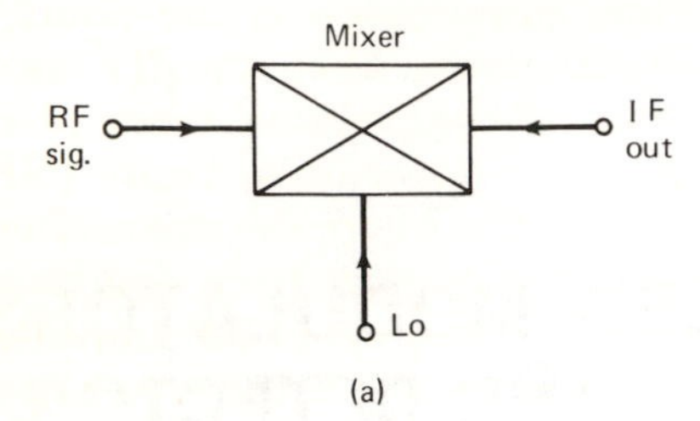

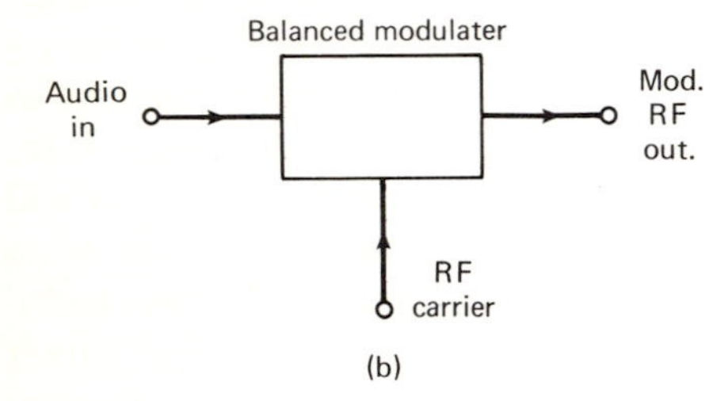

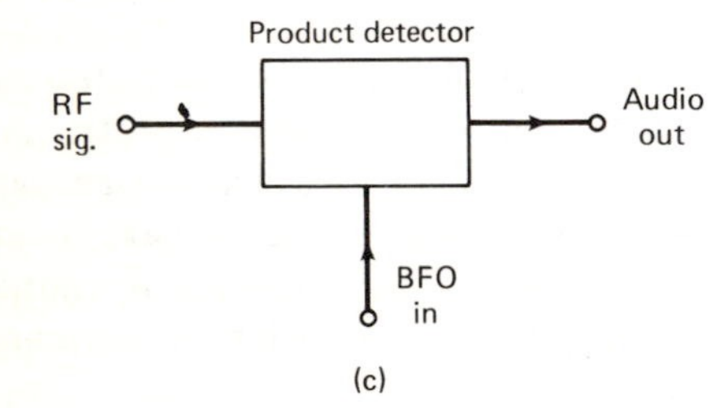

Figure 6.1 Comparison of a mixer, balanced modulator, and product detector.

passive (no operating voltage required) and active (operating voltage needed) devices. Furthermore, we can elect to use the most fundamental form of mixer, the single-ended variety, if high performance is not mandatory, and when equipment cost must be held within specified constraints. The simplest passive mixer is one that contains a single diode. Circuits of this type are sometimes used in UHF and microwave receivers, but are not attractive in circuits that operate below approximately 450 MHz. In the lower VHF region and at high and medium frequencies, it is more desirable to use two diodes in a singly balanced configuration or four diodes in a doubly balanced mixer (DBM). In all instances there will be a conversion loss that is usually of the order of -5 to -8 dB. Additionally, diode passive mixers require substantially greater local-oscillator (LO) power levels to provide the required amount of LO injection to the mixer, as referenced to an active mixer.

Objectively, we want the mixer to withstand strong applied RF signals without creating excessive distortion products. It should also be able to accommodate strong signals without producing cross-modulation effects and gain compression. Gain compression is the point at which a 1-dB reduction in mixer output occurs when the RF input signal level is increased until that value of compression takes place.

Another important consideration is the mixer noise figure (NF). This is especially important above, say, 15 MHz, where the atmospheric noise is often lower in amplitude than is the receiver front-end noise. Mixers are much noisier in mixer service than they are as RF amplifiers when the same device is used in each application. When the mixer noise becomes too great for the operating frequency, we can reduce the effective receiver NF by using a low-noise RF amplifier ahead of the mixer. Its gain must exceed that of the mixer if an improvement is to be realized. Also, the local oscillator and its associated amplifiers must be as spectrally pure as possible in terms of noise before the LO energy is injected into the mixer. Excessive noise output from the LO system will effectively increase the mixer noise figure. Eliminating this noise is an important design problem if we are to strive for high receiver performance. The penalty for using a preamplifier is observed in reduced mixer dynamic range, owing to the increased RF signal level that is applied to the mixer input port. It can be seen from this discussion that all manner of design trade-offs accompany the implementation of a mixer.

Still another important mixer parameter is the intermodulation distortion (IMD) products generated within the device. The IMD should be as low as possible to ensure high dynamic range. IMD is manifested as additional unwanted receiver responses to one or more strong signals in the receiver passband. The effect is one of having a single incoming signal appear in several places as the receiver is tuned away from the signal that is causing the IMD responses. The IMD products are numerically related to the interfering signal and another strong signal nearby in the receiver passband. The IMD characteristics of a given mixer can be determined in laboratory tests by applying two RF signals at the mixer input (two-tone test) and observing the third-order distortion products at the mixer output. A block diagram of a suitable setup for mixer evaluation is shown in Fig. 6.2. A spectrum analyzer, two signal sources, a LO source, step attenuators, a combiner, and post-mixer amplifier and lattice filter are required. The post-mixer amplifier must be as good as or better than the mixer under test (MUT) in terms of IMD. A CATV bipolar transistor (2N5109) meets the requirement easily when operated in a linear broadband configuration.

Port-to-port signal isolation is a vital mixer consideration. That is, we do not want the LO energy to appear at the RF or IF ports of a mixer in sufficient magnitude to create additional frequencies through unwanted mixing action. Doubly balanced mixers offer the best solution to port isolation. Singly balanced mixers are the second choice, and single-ended mixers are the worst we might select for resolving the problem.

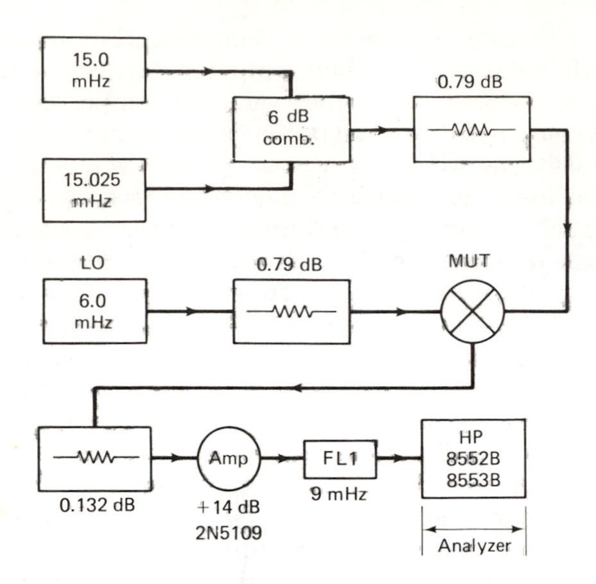

Figure 6.2 Block diagram showing the test setup for measuring mixer dynamic range.

Maximum mixer conversion gain will occur only when the input and output of the mixer are well matched to the characteristic impedances of the source and load. Intentional mismatch is often applied to reduce the conversion gain and to enhance the IMD characteristics by operating the output of an active mixer at low impedance levels. This technique reduces the ac voltage swing and helps to keep the collector or drain of a mixer transistor within a more linear operating range.

6.1.1 Single-Ended Mixers

The practicality of active single-ended mixers is realized in the design of inexpensive receivers, such as the pocket-sized AM band transistor radio. A bipolar-transistor mixer is capable of delivering conversion gains as great as 30 dB under well-matched conditions and careful biasing. Unfortunately, the dynamic range of such a mixer is very poor, which helps to explain the unsatisfactory strong-signal performance of most imported "pocket radios." But this type of mixer contributes to a reduction in receiver stages with respect to overall receiver gain. RF amplifiers are seldom used in receivers of this general variety.

Figure 6.3 contains the circuit of a typical single-ended bipolar mixer. Approximate values are listed for the components and operating voltages for best performance. A mixer is necessarily a nonlinear device and would operate, ideally, with a square-law response. Therefore, the

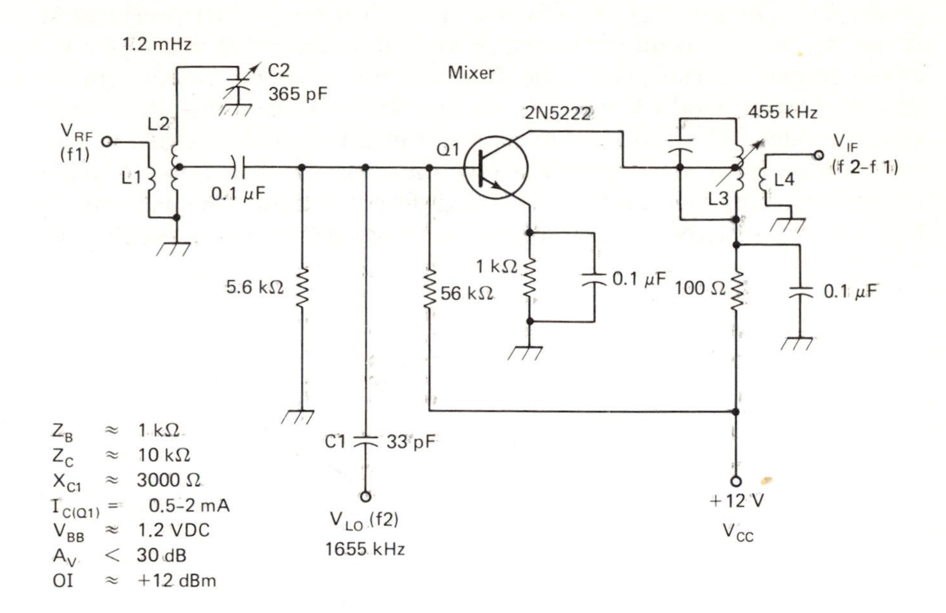

Figure 6.3 Practical circuit for a bipolar-transistor mixer.

forward bias on $Q1$ of Fig. 6.3 is held to a low value. The V_{BB} used will aid conversion gain and enhance the weak-signal response of the mixer. The base of $Q1$ is tapped toward the grounded end of $L2$ to effect an impedance match between the mixer input and the antenna. Typically, the input impedance of this kind of mixer is 1000 Ω.

The collector could be connected directly to the top end of $L3$, since the characteristic impedance of the collector will be on the order of 10,000 Ω with the biasing specified in Fig. 6.3. By tapping the collector down on the tuned circuit we can control the conversion gain (reduce it) to a desired value. Also, as the collector is tapped down on the tuned circuit, we are minimizing the loading effects on the tuned circuit, thereby elevating the Q. This provides increased selectivity at the mixer output. This mixer requires very low injection power from the LO. Since we are applying 0.5 V rms of LO signal across the base-emitter port, the power requirements will be minimal. $C1$ should have fairly high reactance at the signal frequency to prevent undue loading of the input tuned circuit by the LO circuitry. Light coupling via $C1$ will also help reduce oscillator pulling when $C2$ is adjusted.

Mixer distortion (IMD) can be defined either in terms of the IMD products being a specified number of decibels below the desired output signal level (IF) or by means of the third-order output intercept number. The intercept point is an imaginary point established when a 10-dB increase in the two input test tones causes the desired mixer output products to increase by 10 dB, but the third-order IMD products increase by 30 dB. In other words, if we had a mixer that was without a gain-compression trait, we could find a point at which the level of the desired output product would be equivalent to that of the third-order IMD product. We would define this as the third-order intercept point, the position on a set of curves where the desired output-product curve intersected with the curve of the third-order product. This is shown in Fig. 6.4. The curves in this hypothetical example converge at the +20

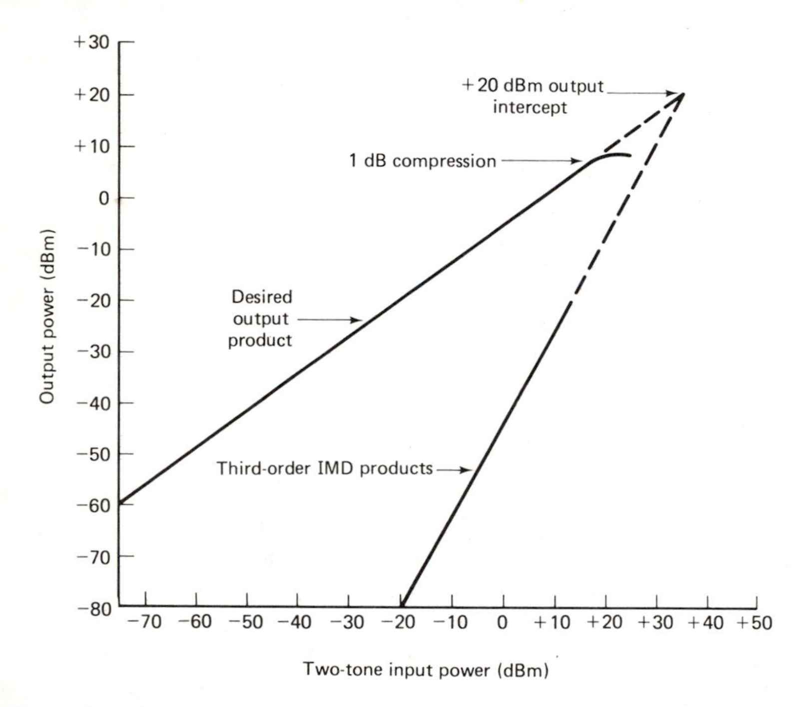

Figure 6.4 Curves that show the third-order output intercept point of a mixer.

dBm output intercept point. It becomes apparent that the greater the output intercept number, the higher is the dynamic range of the mixer. In the case of the circuit of Fig. 6.3 the output intercept is quite low, +12 dBm. Strong mixers yield numbers as great as +45 dBm. A typical number for a medium-quality receiver would be on the order of +25 dBm.

In terms of mixer performance our concern is for the third-order *input intercept* characteristic, even though it is related directly to the output intercept number. The input intercept can be obtained by subtracting the conversion gain of the mixer from the output. Hence, if the conversion gain were 10 dB and the output intercept were +35 dBm, the input intercept would be +25 dBm. It can be seen from this that it is important to minimize the mixer conversion gain in the interest of elevating the input-intercept characteristic. Frequently, this is done in an active mixer by using broadband techniques and introducing intentional mismatches at the input and output ports. Irrespective of the mixer input intercept, the final determination of receiver performance is based on the *receiver input intercept* number. This dictates that care must be given to the design of the input circuitry, inclusive of the RF amplifier that precedes the mixer, in the event that such a stage is used.

A JFET is capable of better performance in a single-ended mixer, compared to a bipolar transistor. Figure 6.5 illustrates how we might

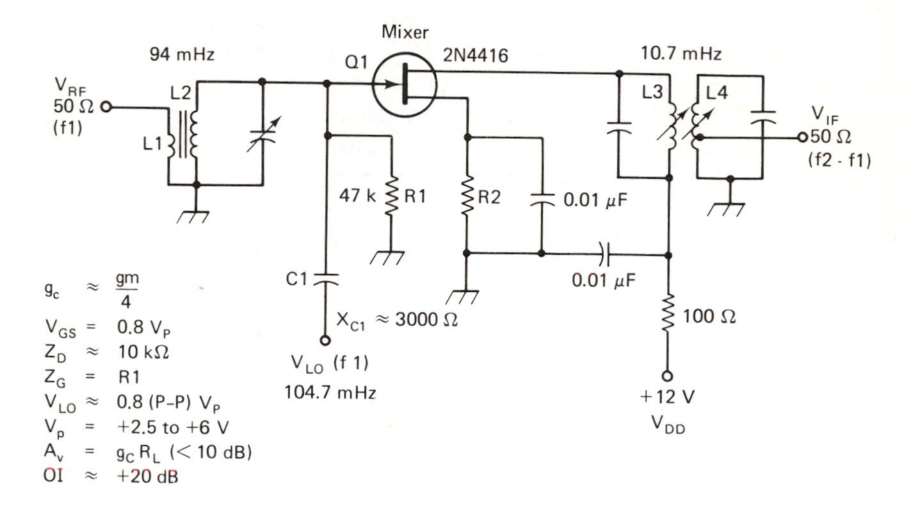

Figure 6.5 Practical JFET single-ended mixer.

use a 2N4416 or similar FET as a mixer. Conversion transconductance (g_c) will be approximately 0.25 the g_m of the same device in amplifier service. Hence, if the FET was rated at 5000 μS as an amplifier, the conversion transconductance would be roughly equal to 5000/4, or 1250 μS. Once we know this number we can approximate the conversion gain by

$$A_V = g_c R_L$$

where A_V is the voltage gain in decibels, g_c is the conversion transconductance in siemens, and R_L is the drain load in ohms. Therefore, if we had a 1250 μS (0.0125 S) g_c and an effective mixer drain load of 1000 Ω, the gain would be

$$A_V = 0.0125 \times 1000 = 12.5 \text{ dB}$$

We will impair mixer performance, however, if we attempt to elevate the conversion gain by increasing the drain load impedance. In the circuit of Fig. 6.5 we would be prudent to tap the FET drain toward the V_{DD} end of $L3$ to force the drain down to a lower impedance than it sees in the circuit, as shown. Typically, the drain impedance will be approximately 10 kΩ with a circuit such as that of Fig. 6.5. This technique will reduce the peak drain-voltage swing, thereby reducing the change in junction capacitance. The latter, which introduces a *varactor effect*, results in the generation of harmonic currents.

Source resistor $R2$ is selected to yield a gate-source dc potential that is 0.8 the pinchoff voltage (V_p) of the FET. In practice, this will be a somewhat arbitrary number, as the manufacturers' data sheets are not specific. V_p is not listed in most of the literature, but we can use the more common parameter, $V_{GS(\mathrm{off})}$, which is of equal magnitude to V_p, but of opposite polarity. For a 2N4416 the spread is listed as -2.5 to -6 V, equating to a V_p of +2.5 to +6 V. Arbitrarily, we can assume a midrange value for V_p, accepting in our design example a voltage of 4.25. Therefore, $R2$ will be chosen to yield a V_{GS} of 0.8×4.25, or 3.4 V. Because the listed $V_{GS(\mathrm{off})}$ values are rather nebulous, we will fare much better in the final analysis by making empirical adjustments to the V_{GS} while the mixer is in operation. A compromise between IMD and conversion gain must be made when establishing the operating conditions. Ideally, the g_c should be held below 10 dB in the interest of good mixer performance.

Under the foregoing guidelines the LO injection should be approximately 0.8 p-p the V_p value. An output intercept of +20 dBm is typical for the mixer of Fig. 6.5, although a theoretical value of +30 dBm exists for FETs such as the U310 in a fully optimized circuit. Since gate injection of the LO is suggested in our circuit example, $C1$ should follow the rule set forth for the circuit in Fig. 6.3. Source injection can be used in this style of circuit, but substantially more LO power will be required and mixer instability may result.

Dual-gate MOSFETs provide comparable performance to JFETs in mixer circuits. They offer the advantage of having a pair of gates rather than a single gate, thus making it easy to introduce the LO and RF signals at separate ports. Normally, the LO energy is applied to gate 2 and the RF input signal is routed to gate 1, as demonstrated in Fig. 6.6. This type of FET is rather fragile in terms of excessive gate voltage, compared to a JFET, but the isolation between the two gates is advantageous toward minimizing LO pulling effects and interaction between the LO output circuitry and the RF signal input circuit. Some manufacturers rate the gate-to-gate isolation as high as 50 dB, but in a practical circuit that magnitude of signal separation would be difficult to achieve. A 25- to 30-dB isolation value is more common, owing to stray coupling external to the MOSFET.

Gate 2 can be biased separately by means of a resistive voltage divider connected to the V_{DD} line. Comparable performance can be ob-

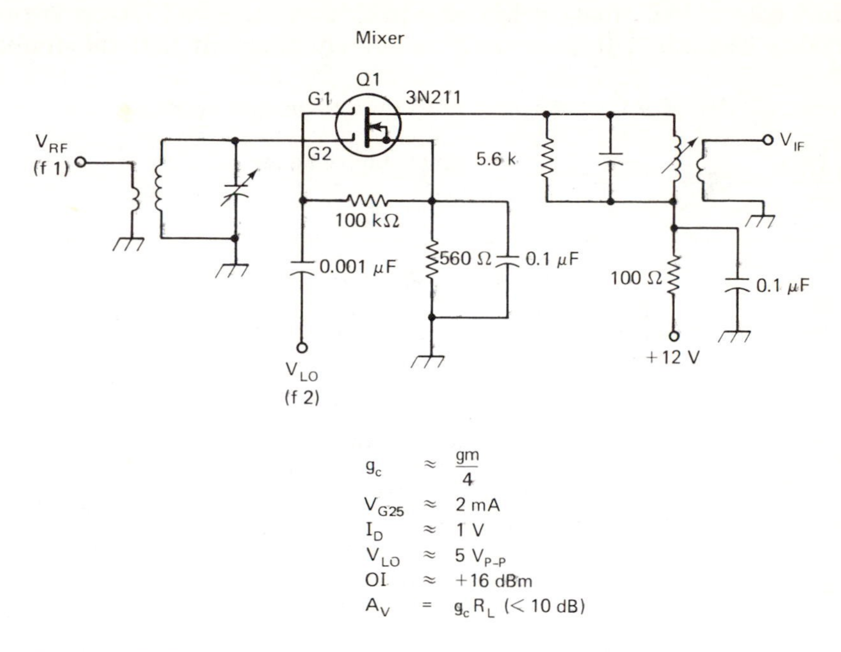

Figure 6.6 Improved single-ended mixer performance can be realized by using a dual-gate MOSFET, which permits two-port injection.

tained, however, by strapping gate 2 to the source through a 100-kΩ resistor, as seen in Fig. 6.6. With a 560-Ω source-bias resistor, an effective gate 2 potential of 1 V will result. Under these conditions, a peak-to-peak LO injection of 5 V will enable the mixer to have reasonable conversion gain and acceptable IMD characteristics. Optimum third-order output intercept for a 3N211 was noted during laboratory tests when V_{g2s} was set at 4.2 V, and with an 8-V p-p LO injection level. This resulted in a third-order output intercept of +16 dBm. Lowering the drain load impedance of *Q*1 in Fig. 6.6 would have aided the output intercept number. The 5.6 kΩ resistor across the output tuned circuit pulls the drain down to an impedance that yields better IMD performance. In a test circuit that contained two 3N211 transistors in a singly balanced mixer, and with broadband coupling transformers at the input and output of the mixer, a third-order output intercept of +20 dBm was observed. The gates and drains were operated at a 200-Ω impedance. Conversion gain was approximately 6 dB when 4.2 V of V_{g2s} was applied, and with an 8-V p-p LO injection level. A 560-Ω resistor was used in each source return.

6.1.2 Balanced Mixers

Balanced mixers offer better port-to-port signal isolation than is possible with single-ended mixers. A comparison of the output products from single-ended, singly balanced and doubly balanced mixers is presented in Table 6.1. It becomes immediately apparent that the doubly

TABLE 6.1 Comparison of Output Products from Mixers

(N)			Single-Ended Mixer		
	f_o	$2f_o$	$3f_o$	$4f_o$	$5f_o$
f_s	$f_o \pm f_s$	$2f_o \pm f_s$	$3f_o \pm f_s$	$4f_s \pm f_s$	$5f_o \pm f_s$
$2f_s$	$2f_s \pm f_o$	$2f_o \pm 2f_s$	$3f_o \pm 2f_s$	$4f_o \pm 2f_s$	$5f_o \pm 2f_s$
$3f_s$	$3f_s \pm f_o$	$3f_s \pm 3f_o$	$3f_o \pm 3f_s$	$4f_o \pm 3f_s$	$5f_o \pm 3f_s$
$4f_s$	$4f_s \pm f_o$	$4f_s \pm 2f_o$	$4f_s \pm 3f_s$	$4f_o \pm 4f_s$	$5f_o \pm 4f_s$
$5f_s$	$5f_s \pm f_o$	$5f_s \pm 2f_o$	$5f_s \pm 3f_o$	$5f_s \pm 4f_o$	$5f_o \pm 5f_s$
(0.5 N)			Singly Balanced Mixer		
	f_o	$2f_o$	$3f_o$	$4f_o$	$5f_o$
f_s	$f_o \pm f_s$	$2f_o \pm f_s$	$3f_o \pm f_s$	$4f_o \pm f_s$	$5f_o \pm f_s$
$2f_s$	—	—	—	—	—
$3f_s$	$3f_s \pm f_o$	$3f_s \pm 2f_o$	$3f_s \pm 3f_o$	$4f_o \pm 3f_s$	$5f_o \pm 3f_s$
$4f_s$	—	—	—	—	—
$5f_s$	$5f_s \pm f_o$	$5f_s \pm 2f_o$	$5f_s \pm 3f_o$	$5f_s \pm 4f_o$	$5f_o \pm 5f_s$
(0.25 N)			Doubly Balanced Mixer		
	f_o	$2f_o$	$3f_o$	$4f_o$	$5f_o$
f_s	$f_o \pm f_s$	—	$3f_o \pm f_s$	—	$5f_o \pm f_s$
$2f_s$	—	—	—	—	—
$3f_s$	$3f_s \pm f_o$	—	$3f_o \pm 3f_s$	—	$5f_o \pm 3f_s$
$4f_s$	—	—	—	—	—
$5f_s$	$5f_s \pm f_o$	—	$5f_s \pm 3f_o$	—	$5f_o \pm 5f_s$

LO = f_o, signal frequency = f_s
$2f_o \pm f_s$ = third-order product
$3f_s \pm 3f_o$ = sixth-order product, etc.
N = number of mixer products

balanced mixer is the stellar performer in terms of isolation and product suppression. It is, thus, the mixer preferred by designers of high-performance equipment.

An example of properly designed JFET singly balanced mixer is seen in Fig. 6.7. It is similar to a circuit suggested by E. Oxner of Siliconix in *Application Note AN72-1*. Dynamic balance is assured through the use of a U430 dual-FET unit. $R1$ is selected to give a V_{gs} of 0.8 the pinchoff voltage, which will be on the order of 2 V, allowing for the published spread of -1 to -4 V, $V_{gs(\text{off})}$. Symmetrical feed is used at the mixer input by means of $T1$, $T2$, and $T3$, providing a 180° phase dif-

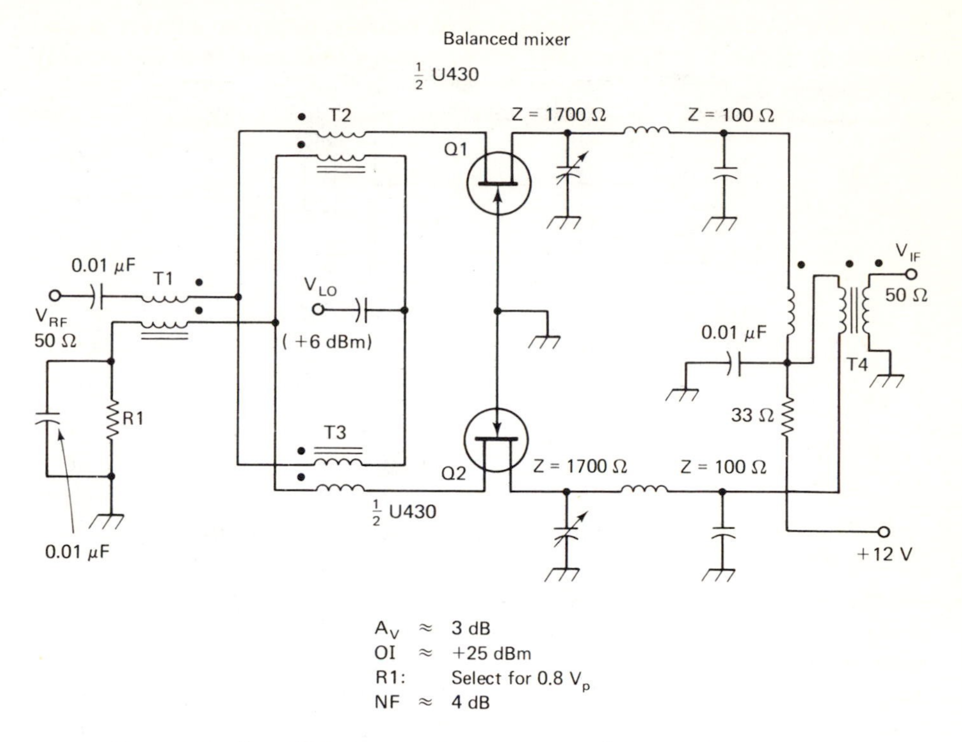

Figure 6.7 **Dual FET balanced mixer using a Siliconix U430 device.**

ferential between the sources of $Q1$ and $Q2$. The transformers are all wound on 125 μ_i ferrite toroid cores, 0.37-in. diameter. $T1$, $T2$, and $T3$ contain bifilar windings of 15 turns of number 26 enameled wire. $T4$ is wound on a 0.5-in. diameter ferrite core (μ_i = 125), trifilar fashion, with 15 turns of number 26 enameled wire. These transformer dimensions are suitable for use from 3 to 30 MHz.

Low-pass matching networks are used in the drains of the FETs to lower the drain impedance to 1700 Ω. They are structured to transform the impedance from 1700 ohms to 100 Ω. The 4 : 1 impedance ratio of $T4$ matches the mixer to a 50-Ω load. Best performance results when V_{LO} is 0.8 p-p $V_{gs(\mathrm{off})}$.

A singly balanced diode mixer is presented in Fig. 6.8. It suffers the same limitations found in singly balanced active mixers, and in addition it yields a conversion loss of approximately 6 to 8 dB. $T1$ is a trifilar-wound toroidal broadband transformer. $T2$ is also a broadband transformer, which is used to provide a step-up transformation to the mixer load. The turns ratio is chosen accordingly. Matched silicon switching diodes are satisfactory in this circuit, but hot-carrier diodes will offer improved performance.

A quad of matched silicon or hot-carrier diodes can be used to

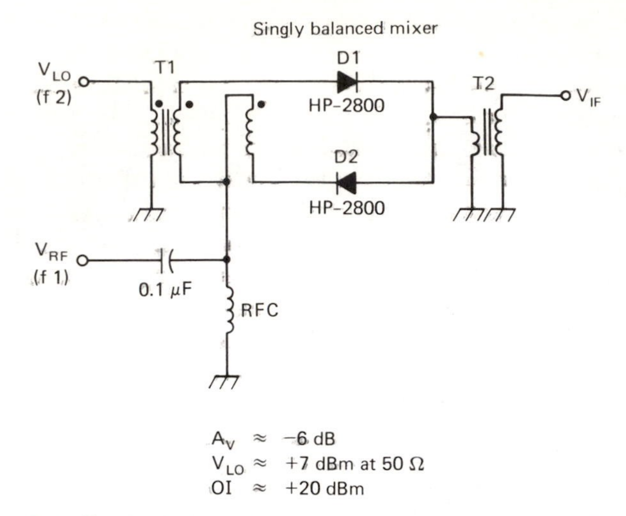

Figure 6.8 Singly balanced two-diode mixer using broadband transformers.

create a doubly balanced diode-ring mixer of the type shown schematically in Fig. 6.9. The circuit, minus *T*1 and *T*4, is typical of the manufactured DBM mixer modules found on the industrial market. Well-designed mixers of this kind offer excellent broadband characteristics, with effective bandwidths of more than 100 MHz. A typical DBM might be rated from 500 kHz to 500 MHz. Medium-level DBMs require an LO power of approximately +7 dBm, whereas the high-level DBMs call for LO powers of +15 dBm or greater. Conversion loss is comparable to that of the mixer in Fig. 6.8, being roughly 6 dB.

Transformers *T*1 and *T*4 of Fig. 6.9 can be added to serve as baluns. This aids the overall balance of the mixer by maintaining sym-

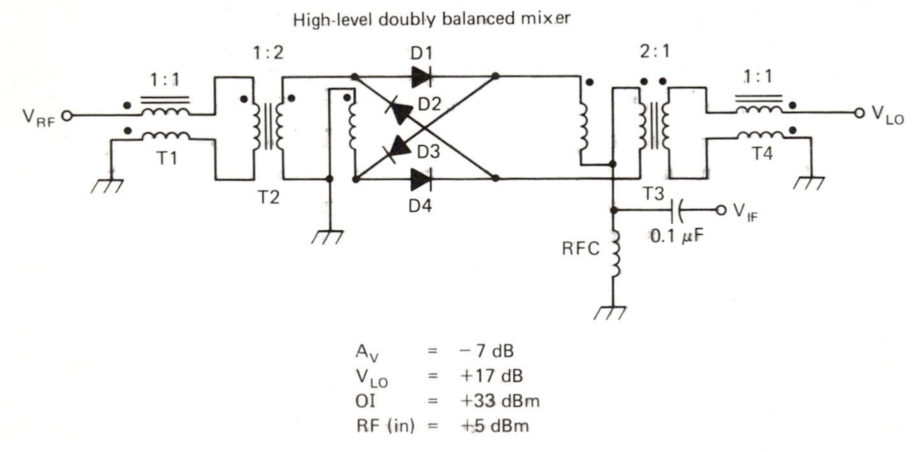

Figure 6.9 High dynamic range and good port-to-port isolation can be had from a doubly balanced diode-ring mixer.

metry at the input and output ports of the circuitry between the input of $T2$ and the output of $T3$. $T1$ and $T2$ are bifilar wound, and trifilar windings are laid on the cores of $T2$ and $T3$. Output intercept for this type of mixer (high level) is typically +33 dBm. Best performance results when the mixer is terminated in 50 Ω by means of a diplexer.

An example of a diplexed DBM is seen in Fig. 6.10. A high-pass network consisting of $L1$, $C1$, and $C2$ is connected to the IF port of the

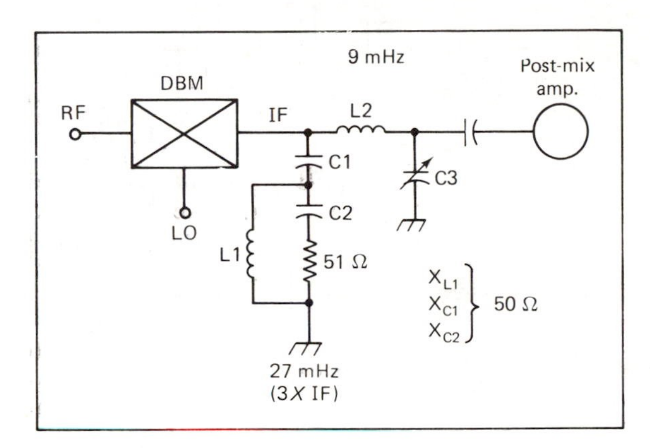

Figure 6.10 A diplexer can be added to a DBM for the purpose of presenting a 50-Ω termination. This improves mixer IMD.

mixer. It is designed for a cutoff frequency of three times the IF. A Q_L of 1 is satisfactory; hence we can use an X_L and X_c of 51 for the resistive termination of 51 Ω. A simple low-pass L network is used between the IF port of the DBM and the post-mixer amplifier to provide an impedance match to 50 Ω. The transformation ratio will be dependent upon the input impedance of the amplifier that succeeds the mixer. Improvements in mixer IMD of 2 to 4 dB have been observed by the addition of the diplexer.

6.1.3 IC Mixers

We could spend countless hours discussing the various types of IC mixers and their respective virtues. Such a treatment is beyond the scope of this text. Therefore, we will examine two of the more common IC mixers for illustrative purposes.

An advantage found in the application of balanced mixers that use ICs is the inherent balance and tracking with temperature that prevails. This is because the bipolar transistors and resistors that are within the IC were all formed at the same time on a common slice of silicon (substrate). This yields transistors with nearly identical dynamic characteristics. Also, when there are temperature excursions of the operating

IC, balance will be maintained through excellent tracking of the component values within the IC package.

A singly balanced IC mixer is depicted in Fig. 6.11(a). It employs an RCA CA3028A IC, which is seen in schematic form in Fig. 6.11(b). The device consists of a differentially arranged pair of *npn* silicon tran-

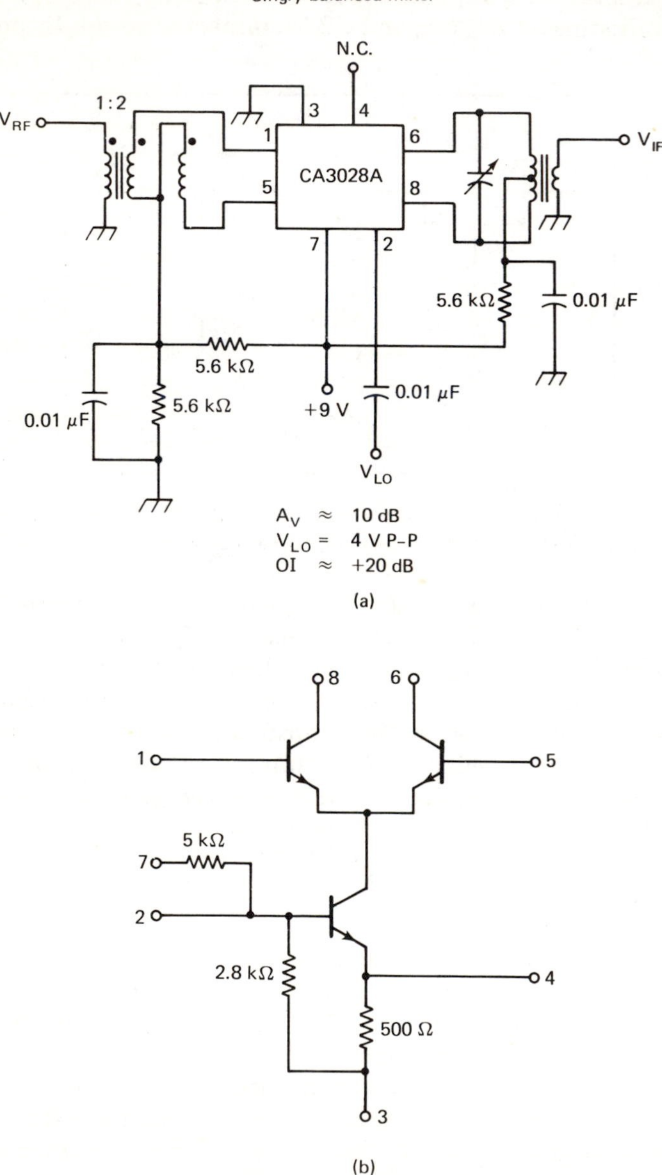

Figure 6.11 CA3028A IC used as a singly balanced mixer; (b) internal circuit of the IC.

sistors. A third transistor functions as a current source in the emitter returns of the differential pair. This IC offers good mixer performance up to at least 100 MHz. As configured in Fig. 6.11(a), it will provide a conversion gain of about +10 dB with an output intercept of approximately +20 dBm. A broadband transformer is used at the input of the mixer to establish a 180° phase difference at the bases of the differential pair. A tuned transformer is used at the mixer output to provide filtering action and a positive conversion gain. The turns ratio of the tuned transformer is chosen to provide a match between the mixer output and the load. The CA3028A has a base-to-base input impedance of approximately 2000 Ω. The collector-to-collector output impedance is on the order of 8000 Ω.

A doubly balanced IC mixer is presented in Fig. 6.12(a). It contains a Motorola MC1496G, which is seen schematically in Fig. 6.12(b). In balanced operation the two differential pairs of *npn* transistors are cross-connected to give a doubly balanced arrangement. Optimum performance will be obtained through experimentation with the biasing values of the circuit. The component values in Fig. 6.12(a) are approximately those suggested by the manufacturer, but the author has seen laboratory reports that indicated improved IMD performance by optimizing the biasing of the IC.

The input and output impedances of the MC1496G are similar to those that characterize the CA3028A of Fig. 6.11. LO suppression at the mixer output is between 40 and 65 dB, depending on the operating frequency and the degree of circuit balance.

6.1.4 High-Level Balanced Mixers

In the interest of high orders of mixer dynamic range, we can utilize mixers that are more powerful than those just described. One IC type of mixer chip that has proved effective in this regard is the Plessey SL6440C. Although similar in concept to the MC1496G, it can be programmed to exhibit a third-order output intercept of +31 dBm. The programmable feature entails the controlled flow of current into pin 11 of the IC. Maximum quiescent current in the differential transistors is 50 mA. Total device dissipation at 25°C is 1200 MW maximum.

Figure 6.13 depicts the laboratory test circuit used to evaluate the SL6440C. Broadband balun transformers are used to provide balanced input and output conditions. This arrangement results in a 50-Ω impedance being presented to the input and output ports of the IC. Conversion gain was measured as 0 dB, although the manufacturer rates it at -1 dB. This number was determined with 50-Ω input and output terminations on the IC (desired to lower the A_V and improve the input intercept) and a programmed current (I_p) of 35 mA. The author measured the noise figure at a signal frequency of 15 MHz and found it

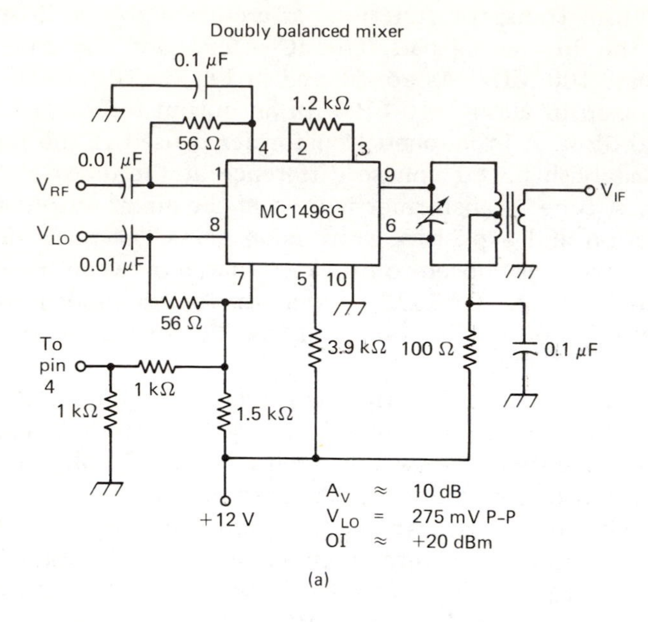

(a)

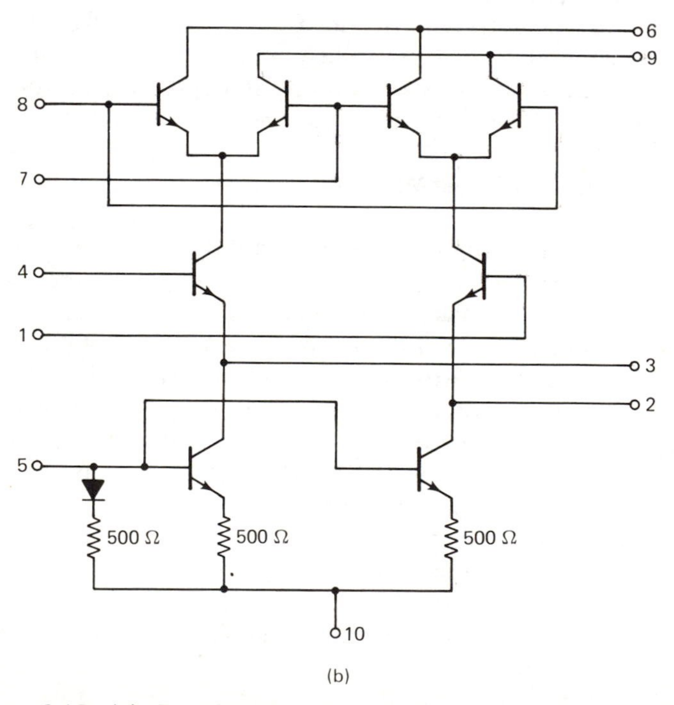

(b)

Figure 6.12 (a) Doubly balanced active mixer with a Motorola MC1496G IC; (b) the inner workings of the chip.

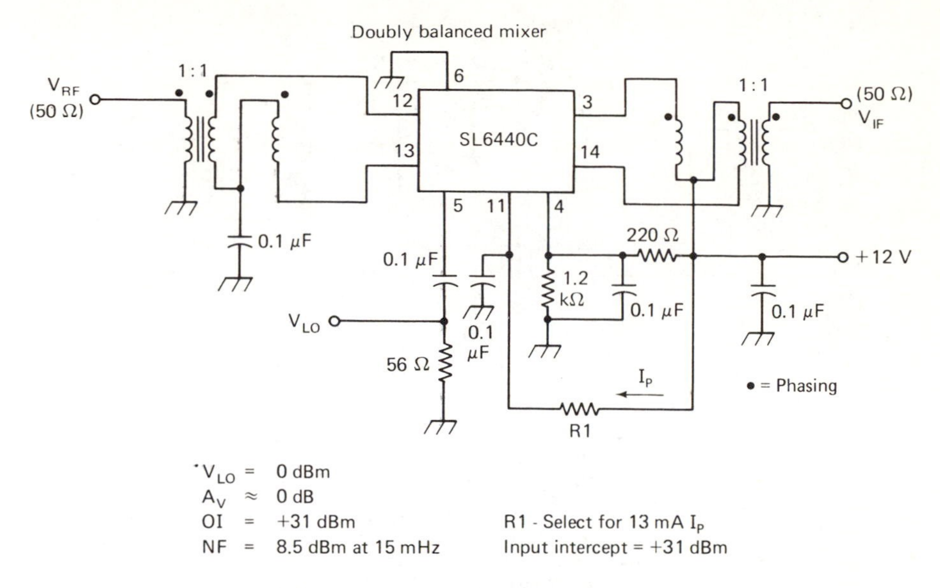

Figure 6.13 A high-level doubly balanced mixer can be obtained by using the Plessey SL6440C IC. Excellent dynamic range is possible with this circuit.

to be 8.5 dB. Plessey rates it at 11 dB at a signal frequency of 100 MHz, the same frequency at which the -1-dB gain was determined.

In the circuit of Fig. 6.13, $R1$ is chosen to permit the desired value of I_p. In the author's tests it was noted that an I_p of 13 mA resulted in optimum mixer performance with an LO level of 0 dBm and an RF signal input of -5 dBm. Under these conditions the LO isolation at the mixer output was 30 dB. The $2f$ LO energy was measured at -75 dB.

If we desire a very strong mixer, we might consider using VMOS power FETs as the active devices. Once again there is the matter of cost-effectiveness to consider, plus the physical bulk of the mixer module. The power FETs are substantially larger than ICs or small-signal devices, and because of the power they dissipate it is necessary to employ heat sinks. Moreover, the power consumption of a VMOS mixer is rather high, placing still another constriction on the design. However, the performance of this variety of mixer offers many advantages, the most significant being high dynamic range if the conversion gain is minimized.

A practical circuit for a high-level, singly balanced VMOS power FET mixer is shown in Fig. 6.14. VHF transistors are used in this laboratory test model, primarily to take advantage of the strip-line packaging format, which lends itself well to heat sinking. Some of the less expensive VMOS units would no doubt yield similar performance, such as the Siliconix VN66AKs, but the TO-5 package would impose restrictions

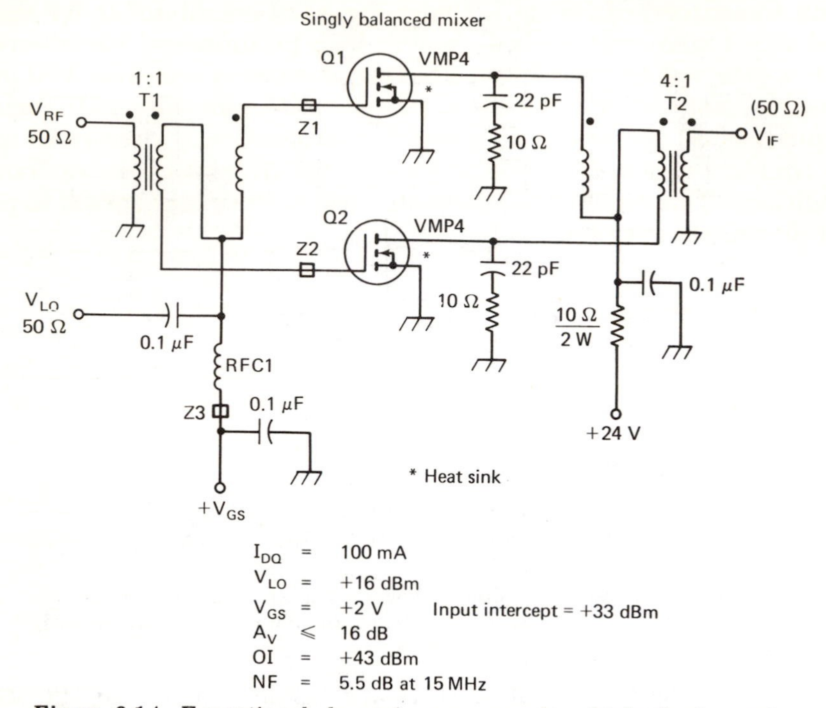

Figure 6.14 Exceptional dynamic range can be obtained when using VMOS power FETs in a singly or doubly balanced high-level mixer. This circuit yields an output intercept of +43 dBm.

with concern to heat sinking of adequate magnitude. The TO-3 equivalent to the VN66AK would probably be a good low-cost alternative to the VMP4s of Fig. 6.14.

In the test circuit we find broadband balun transformers used at the mixer input and output. This forces a 50-Ω gate-to-gate impedance when the generator impedance is 50 Ω. We should keep in mind that, no matter how spectacular the mixer output intercept number, the input intercept is governed by the conversion gain (or loss) of the mixer. Therefore, the mixer gain should be as low as possible, consistent with the design objectives. It is suggested that the *T*1 matching transformer be configured to provide a 50-Ω input impedance (forced through intentional mismatch). We might even wind the transformer to decrease the effective imepdance to some value much lower than 50 Ω. Ideally, the mixer gain should be on the order of unity, thereby yielding an input intercept of +43 dBm. *Z*1 and *Z*2 are necessary in order to damp a VHF parasitic oscillation that became manifest duirng the tests. *Z*3 aids stability by lowering the *Q* of RFC1.

An output intercept of +38 dBm was obtained with a forward gate voltage of 2, an LO injection power of +16 dBm, and a drain current (both transistors) of 160 mA. Under these test conditions an RF signal level of -1 dBm was applied. Similar high performance was observed with a gate voltage of +1.5 and a drain current of less than 100 mA, using the same LO and RF signal powers. LO suppression at the mixer output was 36 dB. The one parameter that may cause concern to us is the conversion gain. It can be as great as +20 dB under some operating conditions. This requires an attenuator pad at the mixer output to protect post-mixer circuitry.

6.2. AM DETECTORS

AM detectors are the least complicated of the various demodulators used in communications circuits. Many receivers, for example, rely on

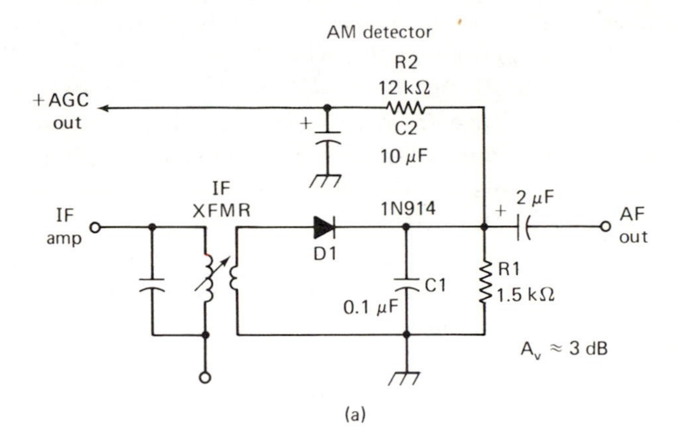

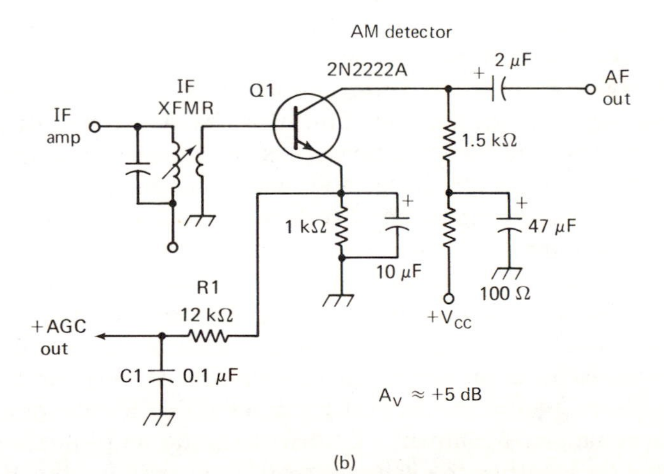

Figure 6.15 (a) One diode is used in this simple AM detector; (b) this circuit utilizes a bipolar transistor as an AM detector to provide a small conversion gain.

a single rectifying diode as an AM detector, and even derive automatic gain control (AGC) from the dc output of the detector diode. A practical example of such a detector is portrayed in Fig. 6.15(a). Operation of $D1$ is the same as that of any rectifier. The IF signal is rectified and the resultant direct current charges $C1$ to a voltage that is proportional to the amplitude of the IF signal and equal to its peak value. When the incoming signal is modulated, the voltage from $D1$ varies in proportion to the modulation (pulsating direct current). Since the pulsations occur at an audio rate, an audio voltage will appear across $R1$, which is selected to represent the approximate impedance of the input of the audio amplifier that follows the detector. For proper performance of the detector to take place, $C1$ should exhibit a low X_c at the intermediate frequency. However, care must be taken not to make $C1$ so great in value that it shunts the high-frequency audio components excessively.

The pulsations in the dc output of $D1$ need to be filtered out of the AGC supply line to prevent the AGC from ramping up and down at an audio rate. A simple RC network ($R2$ and $C2$) provides suitable filtering. Since the AGC voltage from this circuit will be positive, it may be necessary to drive a dc amplifier with it to reverse the effective polarity of the AGC voltage. This will depend upon the kinds of devices used in the receiver RF and IF amplifiers.

The passive detector of Fig. 6.15(a) is capable of high dynamic range and low cost. The limitation is that it is a very lossy detector. The precise amount of power loss will depend upon the input signal level, but a typical range of loss is between 20 and 30 dB. When this is prohibitive to our design objectives, a better choice will be an active AM detector.

If a conversion gain is desired, we can adopt the circuit shown in Fig. 6.15(b). This style of AM detector will provide a higher level of AGC output voltage, along with giving us a small conversion gain. Detection is accommodated via the base-emitter diode junction of $Q1$. The dc and audio components are amplified by the transistor in accordance with the current-gain characteristics of the device chosen. Amplification takes place in the same manner as with any transistor audio amplifier, which means that the output signal from $Q1$ is limited by the collector current and the V_{CC}.

In effect, the base-emitter junction, during rectification, causes a conversion loss, as is true of the diode detector of Fig. 6.15(a). The amplifying action of the transistor does, however, compensate for the loss and ensures an approximate gain of 5 dB. AGC is taken from the emitter of $Q1$. Since it is a positive voltage, its effective polarity may need to be changed by means of subsequent dc amplifier stages. This will depend on whether the gain-controlling circuit for the RF and IF amplifiers requires forward or reverse AGC. AGC filtering is provided by means of $C1$ and $R1$ of Fig. 6.15(b).

6.3 FM RATIO DETECTOR

The two most common varieties of FM detector are the *discriminator* and *ratio detectors*. The former has certain limitations that have made it less popular than the ratio detector. Notably, the discriminator detector, by virtue of being a balanced phase-shift detector, detects AM as well as FM. This results from the circuit being balanced at the IF center frequency. At off-resonance frequencies, variations of amplitude in the IF signal are reproduced in part at the audio output of the detector. This requires the use of a limiter ahead of the discriminator to reduce amplitude peaks.

A practical ratio detector is offered in Fig. 6.16. It responds only to FM signals. The voltages from *D*1 and *D*2 are added, and the sum is made constant by the charging action of *C*1. This relatively constant sum voltage prevents the detector from responding to AM variations in the IF signal, irrespective of the swing above or below the IF center frequency.

The currents through *D*1 and *D*2 are unequal when the IF carrier is displaced from its center frequency. The unequal currents cause a voltage to be developed across *C*2. The cycle runs from positive to negative when this happens, thereby resulting in an audio signal that can be amplified to drive a speaker or headphones. Automatic frequency control (AFC) can be applied to the receiver oscillator by sampling the dc voltage at *C*4. The polarity and magnitude of this voltage will be dependent upon the magnitude and direction of the local oscillator drift.

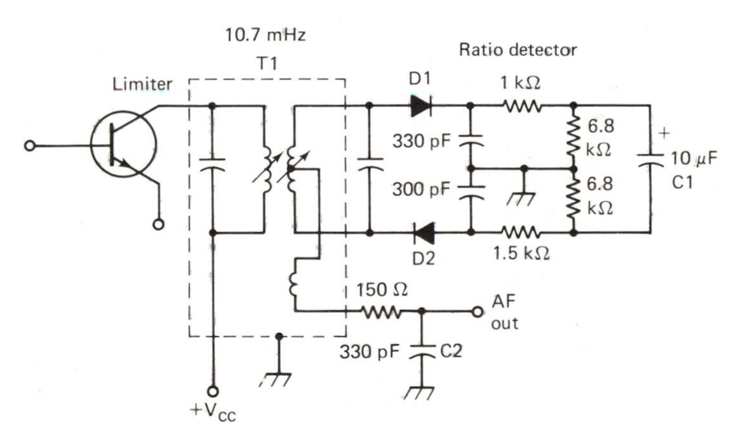

Figure 6.16 Circuit diagram of a practical FM ratio detector for 10.7-MHz operation.

6.4 PRODUCT DETECTORS

Product detectors are what the name implies: the detector output is the product of the two signals that are applied to it, just as we learned in our discussion about mixers at the start of this chapter. The essential difference between mixers and product detectors is that the two RF signals applied to a product detector are on approximately the same frequency, whereas the input frequencies to a mixer are normally somewhat removed from one another spectrally. Product detectors differ from mixers in that they can produce audio two-tone distortion products, second- and third-order sum and difference products, and harmonic distortion. Because of these factors, we will find preferred performance when using a high-level doubly balanced product detector, such as the diode ring that employs four silicon or hot-carrier diodes. This requires a fairly high LO (BFO) injection level, but permits high input signal levels from the IF amplifier chain with minimum difficulties in dynamic range. A balanced modulator also functions similarly to a product detector, but produces an RF output signal, which is the result of mixing an RF signal with an audio signal. The carrier is balanced out in the process. Generally, mixers, balanced modulators, and product detectors operate in the same fashion and with similar dynamic characteristics. Therefore, there will be no special treatment in this text on balanced modulators.

Product detectors are used for single sideband and CW demodulation, although they permit the detection of AM when the signal is tuned to zero beat (no heterodyne present in the receiver output). A beat-frequency oscillator (BFO) generates an injection carrier that is applied to the product detector. The IF signal is routed to the remaining input port of the detector. Output from the detector is at audio, and the output frequency will be dependent upon the frequency difference of the IF and BFO frequencies. Hence, during CW reception, if the BFO is offset 1 kHz from the IF center frequency, a 1-kHz audio note will be heard at the receiver output. In the case of SSB reception, the BFO provides the carrier for the SSB IF signal (which has no carrier), creating an AM type of signal. Detection then follows the lines set forth in the Section 6.2.

6.4.1 MOSFET Detector

Transistors of the dual-gate MOSFET family are often used as simple product detectors. Such devices as the RCA 40673 and 3N211 types are well suited to this service. A practical circuit is presented in Fig. 6.17. The BFO is injected on control-gate 2, which receives its forward bias from the source through a 100K-kΩ resistor. The IF signal is applied to gate 1. The drain voltage varies at an audio rate to produce the desired output from the detector. In this circuit it is important to bypass the source resistor for RF as well as audio. This will assure optimum conversion gain by preventing the FET from being degenerative at audio frequencies.

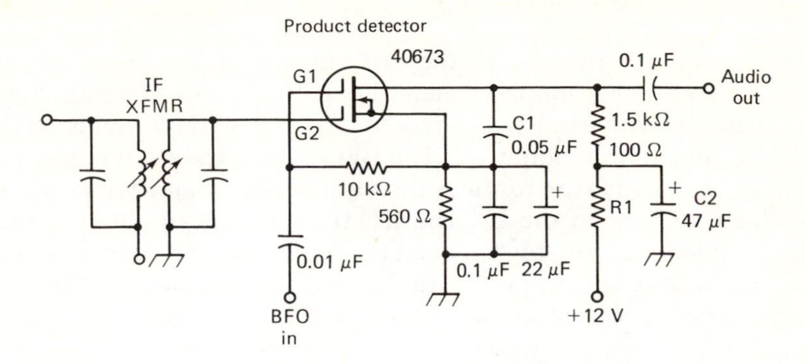

Figure 6.17 A dual-gate MOSFET serves nicely as a simple product detector in this practical circuit.

$C1$ is used to bypass RF energy from the drain circuit, thereby preventing the unwanted energy from being passed along to subsequent audio-amplifier stages. Excessive RF on the audio line can cause instability and desensitization of the audio stage that follows the detector. The reactance of $C1$ must be high enough at audio to prevent shunting the high-frequency audio components to ground. Since the detector of Fig. 6.17 is effectively the first audio stage of the receiver, its V_{DD} line needs to be decoupled from the supply line to the remaining audio stages. If this is not done ($R1$ and $C2$), audio oscillation (motorboating or howling) is likely to occur. Gain with this type of detector is on the order of +10 to +15 dB.

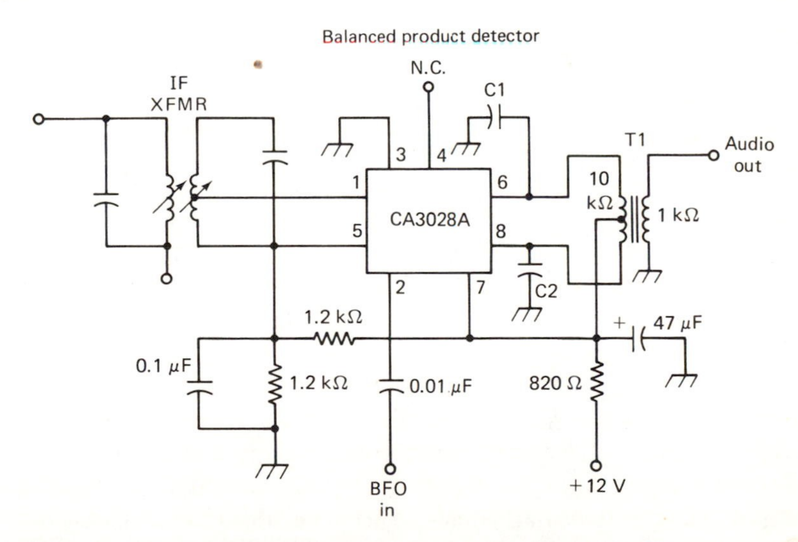

Figure 6.18 A singly balanced product detector can be had by using an RCA CA3028A IC, as shown here.

6.4.2 IC Balanced Detector

Improved suppression of the BFO energy at the detector output can be realized if we adopt a balanced detector for our circuit. A workable circuit of this kind is offered in Fig. 6.18. It contains an RCA CA3028A differential amplifier. The BFO is injected on the base of the current-source transistor to yield the equivalent of emitter injection to the differential pair in the IC. The RF from the IF amplifier is applied in single-ended fashion to the detector bases, but dc balance remains. The audio output is push pull, with $T1$ providing a match to the base of a post-detector audio amplifier. Pins 6 and 8 (collectors) are bypassed to remove BFO energy, which may be present at the detector output. $C1$ and $C2$ will also help to reduce unwanted high-frequency response in communications receivers. They can be selected accordingly. Gain with this product detector will be approximately +15 dB with the circuit values shown. An 820-Ω resistor and a 47-μF capacitor are used to decouple the V_{cc} line to the IC.

6.4.3 Diode-Ring Detector

High dynamic range can be expected if we use the diode-ring product detector seen in Fig. 6.19. $T1$ is used to provide a broadband input circuit. As with the diode-ring mixer described earlier in the chapter, $T1$ is a trifilar-wound transformer. Approximately +7 dBm of BFO power is required for good detector performance. Output from the detector is filtered by $C1$, $C2$, and $RFC1$, all of which should present the proper reactance at RF, but not a reactance that will impair the audio-frequency response. Therefore, $C1$ and $C2$ should have low X_c at RF

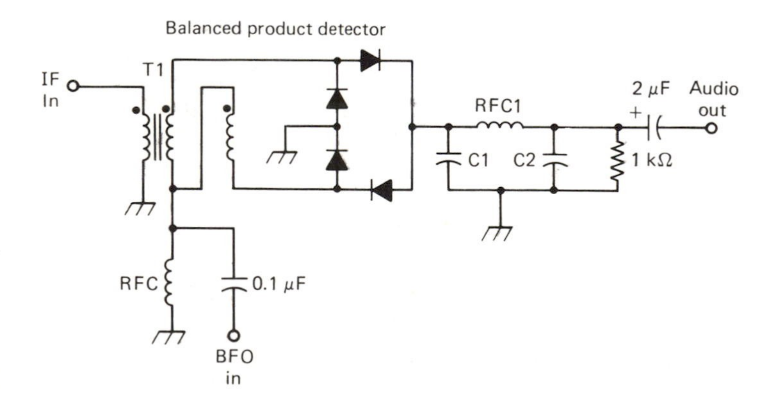

Figure 6.19 A diode-ring product detector of this type is capable of excellent dynamic range, but requires a substantial amount of BFO injection power. Also, it yields a conversion loss.

and high X_c at audio. Conversely, $RFC1$ will present high X_L at RF and low X_L at audio.

As was the case with AM detector of Fig. 6.15(a), there will be a power loss. However, it will not be of the magnitude we encounter when using the single-diode AM detector. Loss with the circuit of Fig. 6.19 will be roughly 6 dB, owing to enhancement brought about by the BFO power.

Balanced diodes are recommended for the diode-ring detector. Matched 1N914s are suitable for most applications. They can be graded out by using a ohmmeter and selecting four diodes that have equivalent forward resistances. Hot-carrier diodes are also excellent for this circuit, and their dynamic similarities are often such that no matching will be necessary. A diode-array IC would offer a good solution to matched diodes and may be worthy of consideration for this and other circuits that require close matching.

6.4.4 Synchrodyne Receiver

For the purpose of illustrating how a product detector is used, let's examine the circuit of Fig. 6.20. There we find an effective but

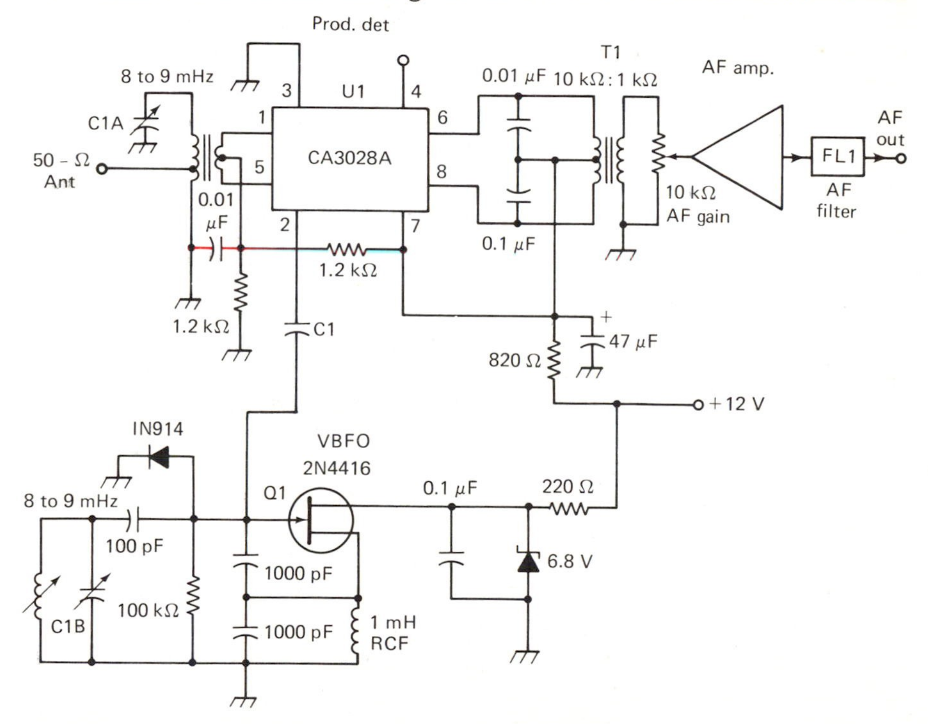

Figure 6.20 Practical circuit of a synchrodyne receiver that uses a product detector at the input. There is no IF strip. Output from the detector goes directly to a high-gain audio amplifier.

very simple receiver arrangement that uses a balanced product detector directly at the antenna input. A variable-frequency beat oscillator (VFBO) is used as the local oscillator and is tunable over the coverage range of the receiver. *C*1, the main tuning capacitor, contains two sections (*C*1*A* and *C*1*B*) to permit tracking and to help disguise VFBO pulling effects during tuning.

Since this kind of receiver (also referred to as a "direct-conversion receiver") has very little front-end gain, and because there is no IF amplifier, a substantial amount of audio gain is necessary. Generally, 75 dB or more of audio gain will be required to accommodate weak signals in terms of ample receiver audio output. A low-noise audio preamplifier can be employed immediately after the detector to enhance the overall signal-to-noise ratio. Selectivity can be obtained simply by including an audio filter, as shown. It can be the passive *LC* type, or an *RC* active bandpass audio filter may be used. The number of filter poles will depend on the skirt selectivity desired.

The primary shortcoming of a synchrodyne receiver is its inability to function as a "single-signal" receiver. That is, there will be a response either side of zero beat, which can at times, depending on band occupancy, cause interference to the signal of interest. The remaining performance feature that can be troublesome is caused by the high degree of audio amplification required. Low levels of hum in the receiver front end can be amplified to annoying proportions. Similarly, if microphonics are present in the front end, they will be quite loud in the receiver output. This condition worsens as the operating frequency is increased.

In an actual situation we would use an RF amplifier ahead of the detector to help prevent the VFBO signal from being radiated by the antenna. The use of a doubly balanced product detector would help to suppress the VFBO energy at the detector input. Additionally, the doubly balanced detector would be less prone to unwanted AM detection than the detector shown in Fig. 6.20. Single-ended and singly balanced product detectors in synchrodyne receivers can be "blanketed" by strong AM signals that are allowed to reach the detector input. A highly selective input tuned circuit will do much to alleviate this problem. Common-mode hum, which can be generated by the VFBO energy entering the power-supply rectifier diodes, will also be unlikely to occur when a doubly balanced detector is used. The Motorola MC1496G IC can be used for the purpose, in place of the CA3028A. Similarly, the diode-ring detector of Fig. 6.19 will offer good performance, but will require greater LO power and approximately 100 dB of post-detector audio amplification.

BIBLIOGRAPHY

Books

Hayward, W., and D. DeMaw, *Solid State Design for the Radio Amateur,* Chaps. 5 and 6, ARRL, Inc., Newington, Conn., 1977.

ARRL, Inc., *The Radio Amateur's Handbook*, receiving chapter, Newington, Conn., 1980, 1981.

Papers and Application Notes

DeMaw D., and G. Collins, "Modern Receiver Mixers for High Dynamic Range," *Preprint* for Session 15, IEEE SOUTHCON El Segundo, CA. 1981. (Also see *QST Magazine*, January 1981.)

Oxner, E., "FETs in Balanced Mixers,"*AN72-1*, Siliconix Inc., Santa Clara, CA. 1979.

Sherwin, J., "FET Biasing," *TA70-2*, Siliconix Inc., Santa Clara, CA. 1979.

Vogel, J., "Nonlinear Distortion and Mixing Processes in FETs," *Proceedings of the IEEE*, vol. 55, no. 12, pp. 2109-2116, New York, NY. 1967.

Will, P., "Reactive Loads—The Big Mixer Menace," *Microwaves*, April 1971, pp. 38-42.

7

IF AMPLIFIERS, FILTERS AND AGC SYSTEMS

Linear small-signal amplifiers differ mainly in their applications. Otherwise, the design approach for IF amplifiers is the same as for an RF amplifier or a video amplifier. The fundamental objectives are, therefore, a prescribed amount of gain, unconditional stability, low signal distortion, an acceptable noise figure (low), and the automatic gain control (AGC) capability needed to provide the desired gain-control characteristics.

Most receivers have AGC systems that ensure a control range of 60 to 120 dB from zero to maximum input signal level at the receiver front end. The AGC threshold point is usually set so that gain reduction commences at some input-signal value below 1 microvolt (μV) at the receiver antenna terminal. Full AGC action takes place (maximum gain leveling) in the region of 10 to 20 μV of input signal. These parameters are of course arbitrary, depending upon the receiver application, circuit complexity, and economic considerations. Broadcast-type AM band receivers, for example, have a very limited AGC range, owing mainly to the relative simplicity of the overall circuit. Many of the "pocket" variety have no RF amplifier and only one IF amplifier, with the AGC being derived by a single diode AM detector. Hence, only one bipolar-transistor stage has AGC, which imposes severe constrictions on the available AGC range. Practically, the AGC range is dependent upon the total gain available from the stages being controlled. Therefore, a well-designed, high-performance communications receiver can easily have an AGC range up to 120 dB. In receivers of this kind, however, it is not uncommon to operate the front-end RF stage without AGC in an effort to preserve the amplifier linearity and noise figure: as the AGC voltage is increased to lower the stage gain, the amplifier linearity is reduced because the bias is changed. A gain-controlled amplifier can operate in class A (no AGC), then degrade to class AB1, class B, and finally to class C with increasing amounts of applied AGC voltage. Optimum linearity occurs in the class A mode. Similarly, the noise figure will change with biasing. Many RF amplifier stages have their optimum noise figures at a specific bias value, and AGC will cause the effective bias to vary with the strength of the incoming signal.

The selectivity of an IF amplifier system is dependent upon the types of *LC* elements used in the circuit. The tuned circuits and their *Q*s are chosen to yield the desired bandpass characteristics, within practical limits. Generally, the *LC* circuits consist of coupling transformers that provide selectivity and a desired impedance transformation between the amplifier stages. Greater selectivity is achieved by inserting IF filters between the receiver mixer or post-mixer amplifier and the first IF amplifier, as shown in the block diagram of Fig. 7.1. Typically,

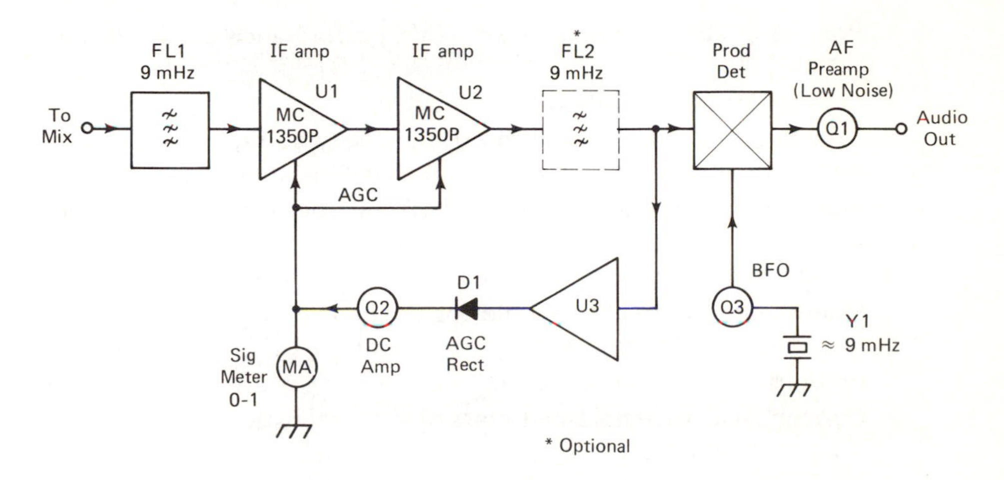

Figure 7.1 Block diagram of an IF strip, filters, product detector, and AGC system. *FL*2 is optional and can be added in the interest of improved skirt selectivity. It will also reduce wide-band receiver noise.

the filter is used as seen at *FL*1. A second IF filter (*FL*2) is sometimes used to improve the shape factor of the amplifier response, which in turn helps to reduce adjacent-channel interference and reject wide-band noise from the IF amplifiers. This form of filter "tail-ending" is found in high-performance HF-band communications receivers.

Figure 7.1 is representative of a typical IF amplifier, detector, and AGC system found in many single-conversion receivers designed for SSB, CW, SSTV, and RTTY reception. Amplifiers *U*1 and *U*2 have a combined gain capability of some 90 dB. IF energy is sampled at the output of $U2$, amplified by means of $U3$, then rectified by $D1$. The resultant dc voltage is applied to a dc amplifier or op-amp, which is tailored to deliver the required AGC voltage swing to control the gain of $U1$ and $U2$. The AGC voltage swing will depend upon the types of active devices employed at $U1$ and $U2$. Some amplifiers (notably ICs) require reduced AGC voltage for maximum gain, while others depend on an increase in AGC voltage for maximum gain. The receiver signal-strength meter (relative) can be operated from the AGC line, as indicated.

We can derive our AGC from the IF strip (RF-derived), or we can sample the audio voltage after the detector and employ *audio-derived AGC*. Again, the choice is made on an arbitrary basis, since there are pros and cons prevalent in the engineering community about each type of AGC.

We might summarize this section with the following criteria suggested to the author by William Sabin, an engineer with Collins Radio Company. He considers the following as the important factors in an AGC system for SSB receivers:

1. Preserving the intermodulation (IM) performance of the desired signal.
2. Preserving out-of-band IM performance.
3. Adjusting the gain distribution dynamically so that the stages are not overloaded as the receiver input signal becomes larger.
4. Maintaining a constant control characteristic in terms of decibels per volt.
5. Good transient response: no overshoots or paralysis.
6. Maintaining a correct threshold for the bandwidth in use.
7. Good filtering of the AGC control voltage to prevent distortion products on the desired signal.
8. Correct AGC threshold and control characteristics of the receiver front-end circuits.
9. Controlling the audio rise for signals between threshold and maximum level.
10. Use amplifiers and AGC methods that do not degrade the IM at higher signal levels. The Fairchild μA757 is an excellent IC for IF amplification and was designed for use with AGC in SSB circuits.

7.1 BIPOLAR AMPLIFIERS

IF amplifiers that use bipolar transistors are found primarily in low-cost entertainment receivers. This is because they are inexpensive, deliver adequate gain for the purpose, and are relatively stable (low gain). We should not accept this as an indictment of bipolar transistors, because it

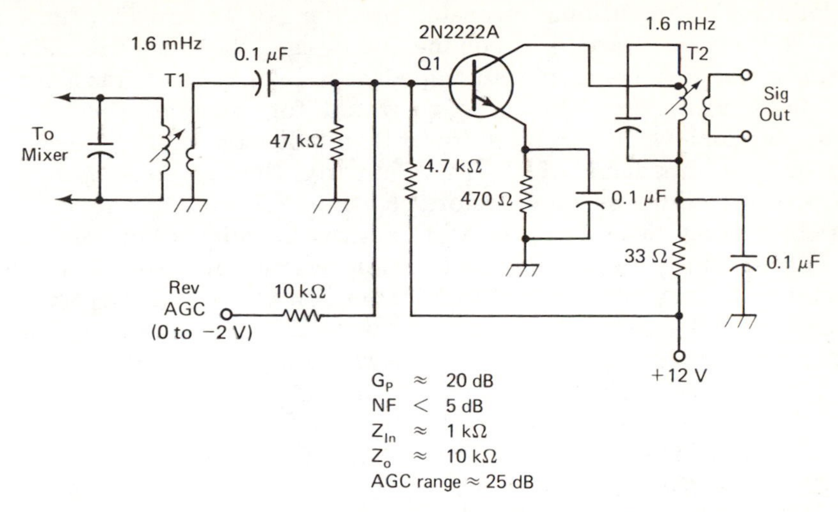

Figure 7.2 Practical circuit for a bipolar-transistor IF amplifier stage. Reverse AGC is applied to the transistor base.

is possible to design a high-performance receiver completely with bipolar transistors: It is important to remember that most linear ICs are simply a large number of bipolar transistors on a common substrate. ICs are preferred by designers of the more sophisticated receivers because they permit a lot of gain in one small "building block," and because they simplify the printed-circuit-board layout with respect to miniaturization.

A typical bipolar-transistor IF amplifier is shown in Fig. 7.2. To ensure ample gain, we should select a transistor with an f_T that is five to ten times higher than the planned operating frequency. The small-signal gain (*hfe*) should be higher than the desired gain of the stage to provide the desired performance. We would choose a device for *Q*1 that has a low noise figure at the specified operating frequency.

Our circuit example shows the use of AGC, which is applied to the base of *Q*1. The amplifier is forward biased for linear operation with no signal present. When a signal appears, AGC voltage is developed elsewhere in the receiver and a control process commences. The greater the incoming signal level, the more negative the AGC voltage becomes, thereby overriding the forward bias of *Q*1. As the forward bias is defeated, the stage gain is reduced. To realize a worthwhile AGC range, three or four stages, such as *Q*1 of Fig. 7.1, would be operated in a cascade configuration. AGC would be applied to all the amplifiers.

Maximum amplifier gain will occur when no reverse voltage is present on the base of *Q*1, and when the input and output of the transistor are matched to the source and load impedances. In multistage amplifiers where the overall gain is high, stability may become a problem (regeneration). Our option in such cases is to trade gain for stability by introducing an intentional mismatch by means of *T*1 and *T*2 (Fig. 7.1). This is common practice in commercial design. This measure calls for lowering the input and output impedances seen by the transistor. In creating a mismatch of this type, we would, for example, adjust the *T*1 turns ratio for a value somewhat lower than that required for the characteristic base impedance of 1000 Ω. Similarly, the tap point on the primary of *T*2 could be moved closer to the 12-V end of the winding, thereby pulling the collector down to a lower impedance. In the process of mismatching, the *Q*s of the transformers will increase to provide greater selectivity through reduced loading on the resonators. We must assume in this exercise that we know what the source and load impedances (mixer output and second IF amplifier input) are.

The usual small-signal RF or IF amplifier of the kind we see in Fig. 7.1 has an input impedance of 500 to 1500 Ω and a collector impedance on the order of 10,000 Ω. An AGC range of approximately 25 dB is possible with this circuit.

7.2 MOSFET IF AMPLIFIER

A more suitable choice of device for an IF amplifier that uses discrete components is a dual-gate MOSFET. The dynamic range of a field-effect transistor generally exceeds that of a small-signal bipolar transistor in linear amplification. Also, the input impedance of a FET is very high when compared to the low input impedance of a bipolar device. The higher input impedance (usually 1 MΩ or greater) simplifies the matching procedure. A power gain of up to 30 dB is possible with a dual-gate MOSFET, owing to the high forward transconductance of such devices as the RCA 40673 and the 3N211 MOSFET. A g_m of 17,000 to 40,000 μS is typical for a 3N211. An AGC range of up to 55 dB is not unusual for a 3N211 transistor, making it ideal for use in an IF amplifier system.

A practical circuit that uses a 3N211 transistor is seen in Fig. 7.3. To obtain the maximum AGC range possible with this MOSFET, we must swing the control gate (*G*2) approximately 2 V negative. The requirement can be satisfied readily by referencing gate 1 to the source and elevating the source above ground as shown. An LED is used in our example to develop 1.5 V of source bias, which is ample for good AGC range. A silicon diode could be placed in series with *D*1 to provide a reference of approximately 2.2 V for full AGC action of *Q*1. Under maximum signal conditions, gate 2 of this circuit is effectively biased at -1.5 V. A suitable low-voltage Zener diode can be used in place of the LED at *D*1. The LED barrier voltage is 1.5; hence it serves well as an inexpensive reference element.

Most dual-gate MOSFETs operate efficiently into the UHF region, which can lead to VHF and UHF parasitic oscillations if care is not

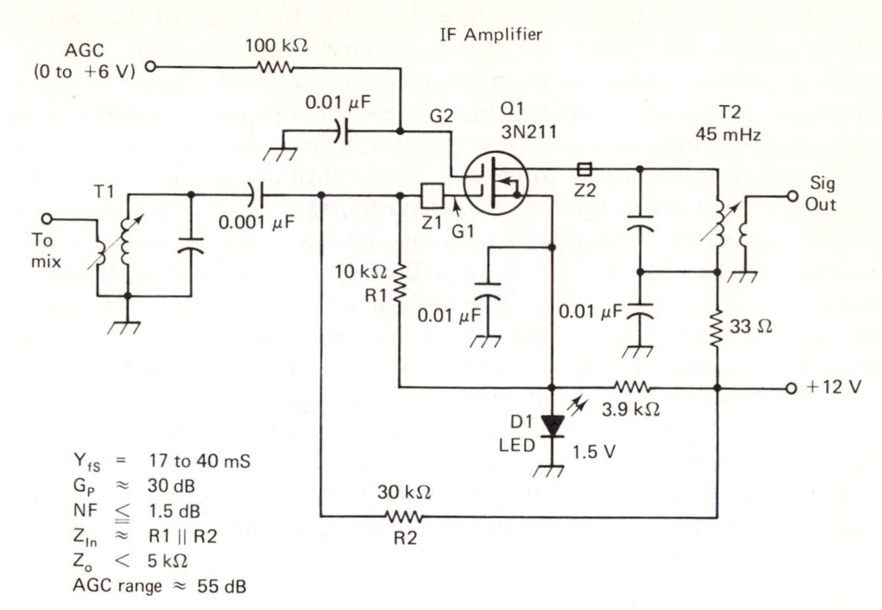

Figure 7.3 An AGC range up to 55 dB is possible with this practical circuit, which uses a dual-gate MOSFET as an IF amplifier.

taken in the layout of the amplifier. When this becomes a problem, it can be resolved easily by employing a medium-permeability ferrite bead at the drain, at gate 1, or both. $Z1$ and $Z2$ of Fig. 7.3 are included as parasitic suppressors. A μ_i of 40 or 125 is suggested for the beads used as suppressors.

Instability at or near the intermediate frequency can be treated in the same manner described for the circuit of Fig. 7.2. Again, intentional mismatch may be necessary to prevent regeneration. This would require that we tap gate 1 of $Q1$ toward the ground end of the $T1$ secondary. The $Q1$ drain could be tapped down on the primary of $T2$ to enhance stability.

Although the characteristic input impedance of a FET is in the megohm region, the actual terminal impedance will be dictated by the gate resistors used. Thus, the equivalent parallel resistance of $R1$ and $R2$ of Fig. 7.3 will constitute the impedance seen by $T1$, 7500 Ω.

7.3 INTEGRATED-CIRCUIT IF AMPLIFIERS

We mentioned earlier that ICs are devices that contain many bipolar transistors on a single slice of silicon material. The type of ICs we might utilize for IF amplification are generally configured to function as differential amplifiers, although many of them can be connected externally for use as cascode amplifiers. The differential format is the most common of the two. Stability is good, despite the high gain of the devices, typically 40 to 60 dB per IC. This is because the connecting leads between the various transistors in the chip are micro-short, so to speak. Care must be taken when making the external circuit connections to eliminate long leads and potential ground loops. This, plus good input-output isolation, will help prevent self-oscillations. Similarly, we need to provide each stage with an *RC* decoupling network to prevent unwanted migration of energy from stage to stage via the V_{cc} line. ICs that are suitable for use as IF amplifiers have separate bias terminals that make them readily adaptable to being used with AGC. If AGC is not desired, the bias terminals can be programmed for a specific amplifier gain by means of simple resistive dividers connected to the V_{cc} line.

Most linear ICs are available in two package sytles, TO-5 (circular) or the DIP (rectangular, dual-in-line plastic). Some designers prefer the DIP style of package because it is somewhat easier to lay out during pc-board planning. Also, the DIP package will, in some instances, provide better stability because of input-output isolation improvement over the TO-5 package format.

7.3.1 CA3028A Amplifier

An IC that lends itself well to cascode or differential applications is the RCA CA3028A. Since the rated gain is on the order of 40 dB maximum, stability is easy to achieve with proper pc-board layout. Internally, it consists of a differential pair of transistors, the emitters of which are common and are returned to ground through a single bipolar transistor that functions as a current source. The bases and collectors of the differential pair are connected to separate pins, thereby enabling the designer to use the chip in a differential or cascode arrangement. Also, the amplifier can be used in push-pull or single endedly, depending on the application. The current source has internal biasing resistors, but those can be ignored if desired by connecting to the pins that are directly common to the transistor base and emitter. The CA3028A works well as a singly balanced mixer or push-push frequency doubler, with respect to nonamplifier circuitry.

Figure 7.4 shows a practical circuit that uses the CA3028A. Two of these stages in cascade are sufficient for most receiver IF amplifier strips. The characteristic base-to-base input impedance of the IC is approximately 2000 Ω, with a collector-to-collector characteristic of some 8000 Ω. The AGC range for one IC is close to 60 dB. AGC is applied to bias pin 7 of the IC. Maximum gain occurs with 9 V on pin 7, and minimum gain will be observed when there is a 2-V or lower potential at pin 7.

A transfer curve for the CA3028A is shown in Fig. 7.5. We can observe that the linearity of the IC is excellent with signal voltages up to

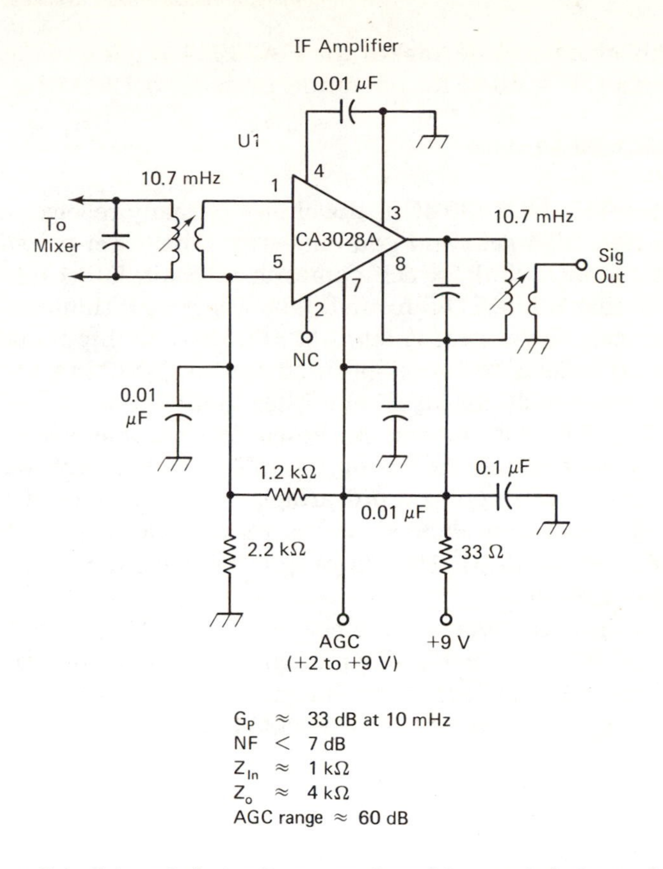

$G_P \approx$ 33 dB at 10 mHz
NF $<$ 7 dB
$Z_{In} \approx$ 1 kΩ
$Z_o \approx$ 4 kΩ
AGC range $\approx$ 60 dB

Figure 7.4 Integrated circuits are preferred by most designers for use as IF amplifiers. A single IC such as the CA3028A can provide high gain and wide AGC range at low cost.

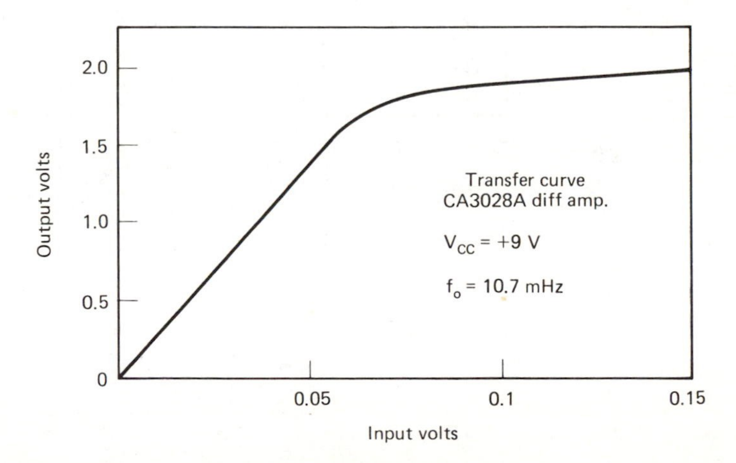

Figure 7.5 Transfer curve for a CA3028A IC showing a linear response with signal-input levels up to 60 mV.

60 mV. This characteristic makes the CA3028A a good choice for IF amplifier service. It is rated for operation from dc to 120 MHz.

7.3.2 MC1590 Amplifier

The Motorola MC1590 IC is the choice of many receiver designers because it offers high gain and good linearity. The lower-priced equivalent part is the MC1350P, which comes in an 8-pin DIP package. The MC1590 is available in a TO-5 format. Operating conditions for the two ICs are identical. Power gain is rated at +60 dB. The input and output impedances are similar to those specified for the CA3028A of Fig. 7.4.

A practical circuit for an IF amplifier that contains an MC1590 is offered in Fig. 7.6. Although it is shown as a single-ended amplifier, pins 1 and 3 can be connected for push-pull input. Similarly, we can use pins 5 and 6 for a push-pull output arrangement. The circuit shown in Fig. 7.6 is the least complicated and is the choice of most designers. The noise figure for the MC1590 is given in tabular form at the right of the schematic diagram.

AGC voltage is applied to pin 2 of the MC1350P (pin-out not shown for the MC1590 in Fig. 7.6). Maximum gain with this IC takes place when the AGC voltage is the lowest (3 V). Hence, minimum gain will be obtained when there is a +6-V potential on pin 2 of the IC. An

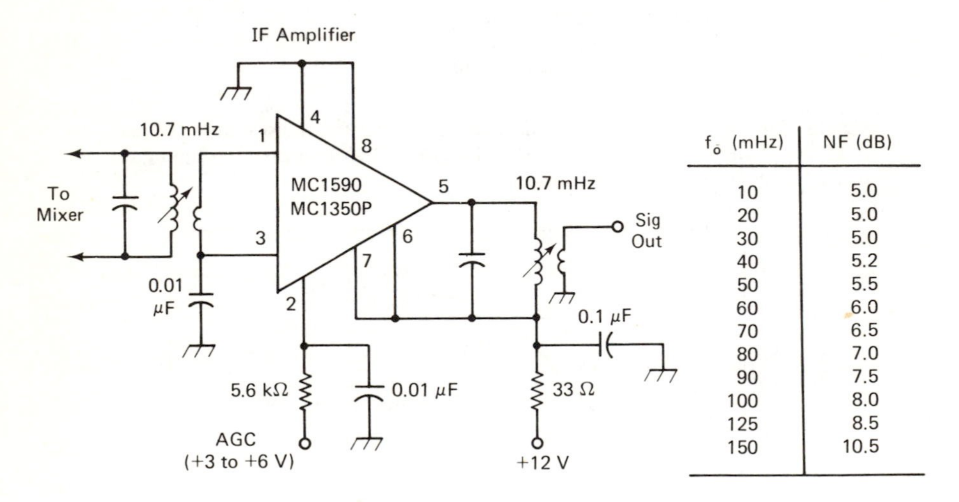

f_o (mHz)	NF (dB)
10	5.0
20	5.0
30	5.0
40	5.2
50	5.5
60	6.0
70	6.5
80	7.0
90	7.5
100	8.0
125	8.5
150	10.5

$G_P \approx$ 60 dB at 10 mHz
NF $\leq$ 5 dB
$Z_{In} \approx$ 1 kΩ
$Z_o \approx$ 5 kΩ
AGC range $\approx$ 70 dB

Figure 7.6 Practical circuit for a 60-dB IF amplifier using a Motorola MC1590 or MC1350P IC.

AGC range of approximately 70 dB is possible with the MC1590/MC 1350P.

Normally, we would use only two MC1590s in a receiver IF amplifier. This is shown in Fig. 7.7. A power gain of roughly 100 dB can

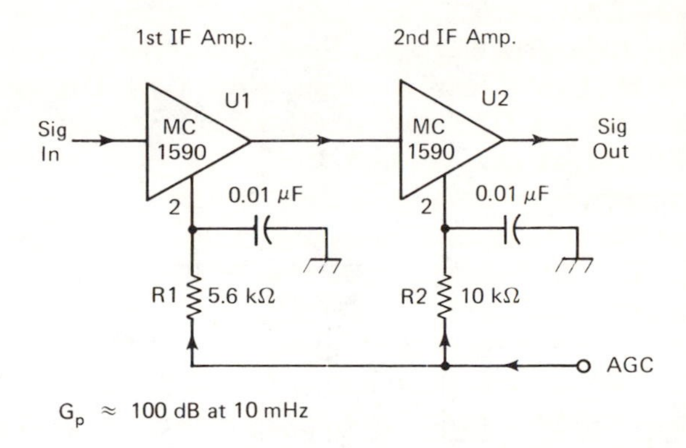

Figure 7.7 Two MC1590 ICs deliver 100 dB of gain with this arrangement. *U*1 is biased to prevent overloading of *U*2 (*R*1).

be obtained from two MC1590s in cascade. The values for *R*1 and *R*2 are recommended by the manufacturer to provide an efficient AGC action. Essentially, the 5.6-kΩ resistor causes the AGC action of *U*1 to be faster than that of *U*2. This prevents overloading of the second-stage input in the presence of large signals.

7.3.3 Composite IF System

An IF system that contains a product detector and an AGC circuit is depicted schematically in Fig. 7.8. In this practical circuit we find Plessey SL1612 ICs serving as IF amplifiers. The net power gain is on the order of 70 dB, with an AGC range in excess of 100 dB. An operating frequency of 9 MHz is specified in this example.

Capacitive coupling is used between the stages, which simplifies the design considerably. Selectivity is dependent upon the characteristics of the IF filter that precedes *U*1.

An SL1641 is used as a balanced product detector (*U*3). A BFO injection level of 75 to 100 mV is required for best detector action. *RC* decoupling immediately follows the detector to prevent BFO energy from reaching the audio amplifiers. *Q*1 functions as a single-pole audio filter to reduce wide-band noise and shape the audio passband for SSB communications.

A 741 op-amp (*U*4) boosts the audio level some 18 dB, after which it is split into two channels, one for the main audio amplifier in the receiver and the other for driving *U*5, an SL1621, which provides

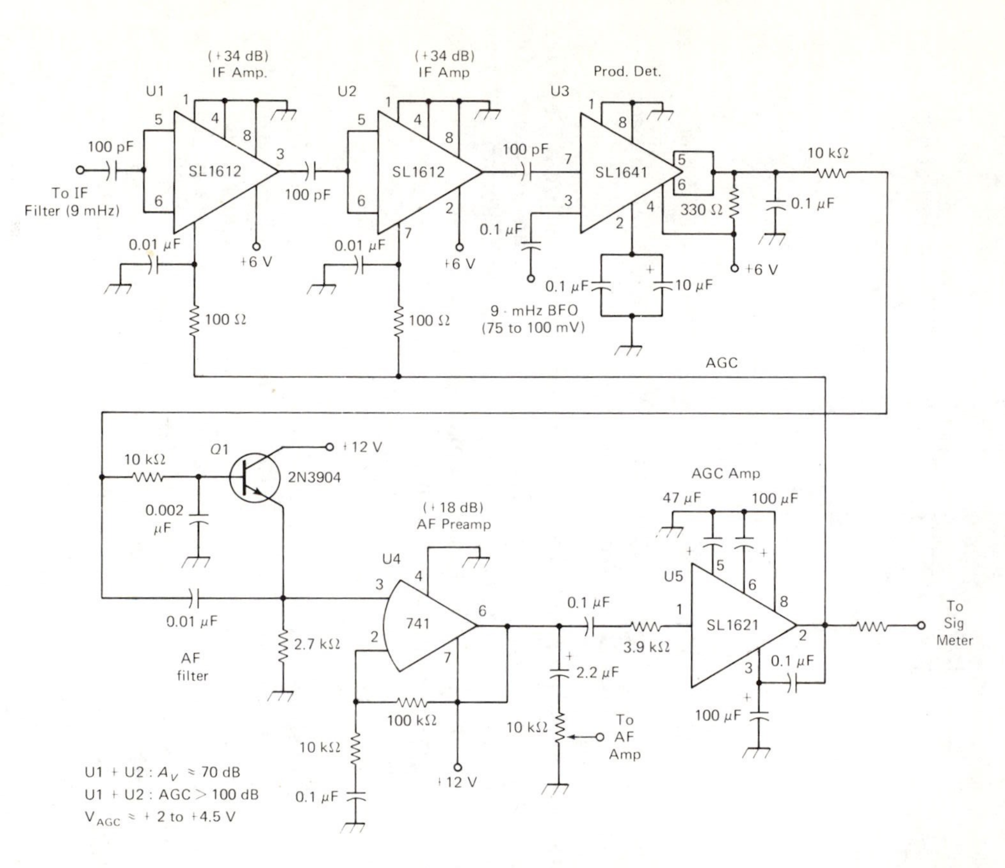

Figure 7.8 Complete practical IF amplifier, product detector, and audio-derived AGC system using Plessey ICs.

audio-derived AGC for $U1$ and $U2$. The AGC voltage range is +2 to +4.5 V. An S meter (signal meter) can be connected to the AGC line if desired.

7.3.4 Frequency Modulation IF Amplifier

The essential difference between an IF amplifier for FM reception as compared to one for AM, SSB, and CW reception is the lack of AGC. A practical circuit for an FM receiver IF system is shown in Fig. 7.9. MC1350P ICs are used as IF amplifiers at $U1$ and $U2$. They are biased for maximum gain by means of resistive dividers at pin 5.

A third MC1350P functions as a limiter at $U3$. The purpose of the limiter is to suppress amplitude peaks of variations that accompany the desired FM signal. $U3$ saturates on AM peaks, thus serving as a limiter. In effect, it operates as a third IF amplifier, which, without AGC on $U1$ and $U2$, becomes an overdriven amplifier with respect to amplitude peaks, such as short-duration noise pulses. The tuned transformers specified in Fig. 7.9 help to ensure maximum gain in the system, while offering IF selectivity.

7.4 IF FILTERING

One of our design objectives when developing an IF system for a receiver or SSB transmitter is to provide a specified degree of overall selectivity. For most modern communications circuits we would employ an IF filter rather than the lumped LC resonators that were common in days past. High-Q LC filters are practical at very low frequencies for some forms of communications, since good selectivity can be obtained at, say, 50 or 100 kHz with LC filters. However, the availability of mechanical, quartz, and piezo monolithic filters makes it more convenient to use them in our designs.

We must take into account a number of parameters when selecting a filter, such as the intermediate frequency, filter insertion loss, terminal impedance, bandpass characteristics, and shape factor. A graphic presentation of some of these parameters is shown in Fig. 7.10. The filter chosen should have good ultimate and spurious attenuation to prevent unwanted responses either side of the filter center frequency (f_c). Also, minimum passband ripple is desired if we are to observe uniform amplitude of the signal across the nose of the filter response curve. The deeper the dips (ripple), the more pronounced is the effect when tuning across a signal.

We are especially interested in the filter shape factor, because it determines the "skirt selectivity" of the IF system. We might consider a shape factor of 1 as the ideal, but skirts that are the same width as the

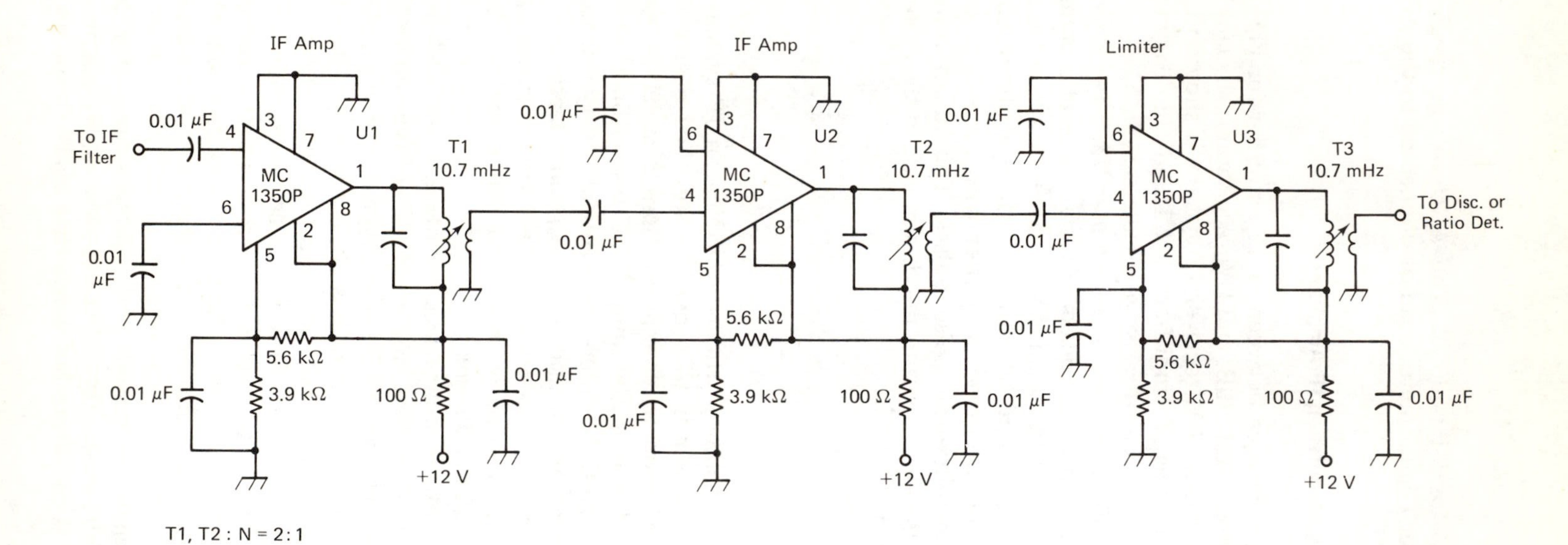

Figure 7.9 An FM type of IF system that includes a limiter. No AGC is used in this circuit.

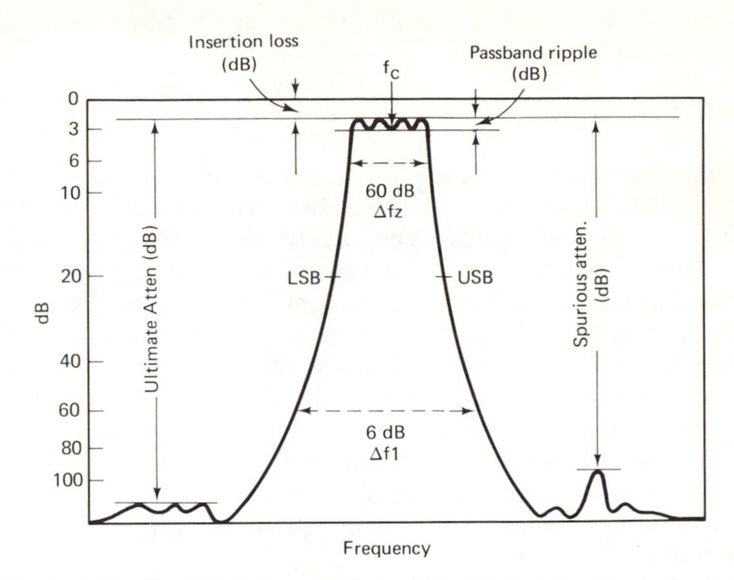

Figure 7.10 Graphical presentation of the bandpass response for an IF filter. Parameters are discussed in the text.

nose of the response curve are impractical. The shape factor is defined as the ratio between the frequency difference at two power points on the response curve. Generally, this measurement would be taken at the 3- and 60-dB points on the curve. Some designers prefer to use the 6- and 60-dB power points (see Fig. 7.10). By way of example, if the 3-dB bandwidth of a filter was 2 kHz and the 60-dB bandwidth was 8 kHz, our shape factor would be 4. The lower the number, the better is the rejection of adjacent-channel interference during reception.

Filter bandwidths at the 3-dB points are in the range of 3 to 8 kHz for AM reception, depending upon the fidelity desired. Most SSB filters have a bandwidth of 2.1 to 2.7 kHz. For CW reception the chosen bandwidths usually fall between 200 and 600 Hz at the 3-dB points. The naturalness of the voice energy is affected by the filter bandwidth and the placement of the BFO frequency on the filter response curve. Most SSB receivers inject the BFO signal at a frequency that falls approximately 20 dB down on either side of the IF curve, as indicated in Fig. 7.10. The BFO offset for CW reception is on the order of 700 Hz, which places it higher on the filter curve. Therefore, if the CW filter f_c was 9.0 MHz, the BFO would operate at either 9000.7 kHz or 8999.3 kHz. The displacement for SSB reception is usually between 1.3 and 1.5 kHz from f_c. Bandwidths for FM voice communications are determined by the deviation amount used in the FM system, such as land-mobile equipment in the VHF and UHF spectrums. The filter bandwidth for 5-kHz deviation from the transmitter would be approximately 10 kHz (receiver filter). It follows that, if the deviation standard was 15 mHz for the transmitter, the receiver IF filter would have a bandwidth of 30 kHz.

7.4.1 Crystal Filters

Quartz crystals lend themselves nicely to the fabrication of bandpass filters. This is because they are stable and exhibit high magnitudes of Q. Since the crystals in this type of filter are of the fundamental-cut variety, the useful upper limit for crystal filters is approximately 20 MHz. Piezo filters of the monolithic kind are used below as well as above 20 MHz, and can be fabricated for use in the VHF spectrum.

The greater the number of crystals or "poles" in a crystal filter, the better is the shape factor. Perhaps the simplest effective crystal IF filter we might employ is the half-lattice version seen in Fig. 7.11. It would be suitable for a low- or medium-performance receiver, but lacks the skirt selectivity needed for most communications work today. Insertion loss with a properly designed half-lattice filter is low, usually less than 5 dB.

A basic half-lattice with two shunt crystals offers somewhat better skirt selectivity, as indicated in Fig. 7.12. $C1$ is used to compensate for the holder capacitance and is adjusted for the best bandpass uniformity. $C1$ may be required in parallel with $Y2$, as shown, or it may be needed

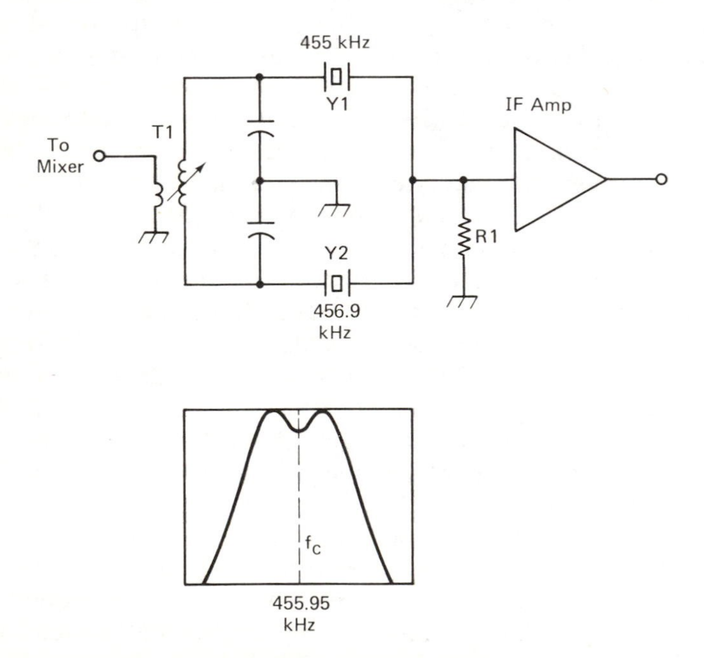

Figure 7.11 One of the simplest bandpass IF filters in use is the half-lattice type shown here. $Y1$ and $Y2$ are selected to provide the desired nose selectivity, but the skirt selectivity will be poor with this filter.

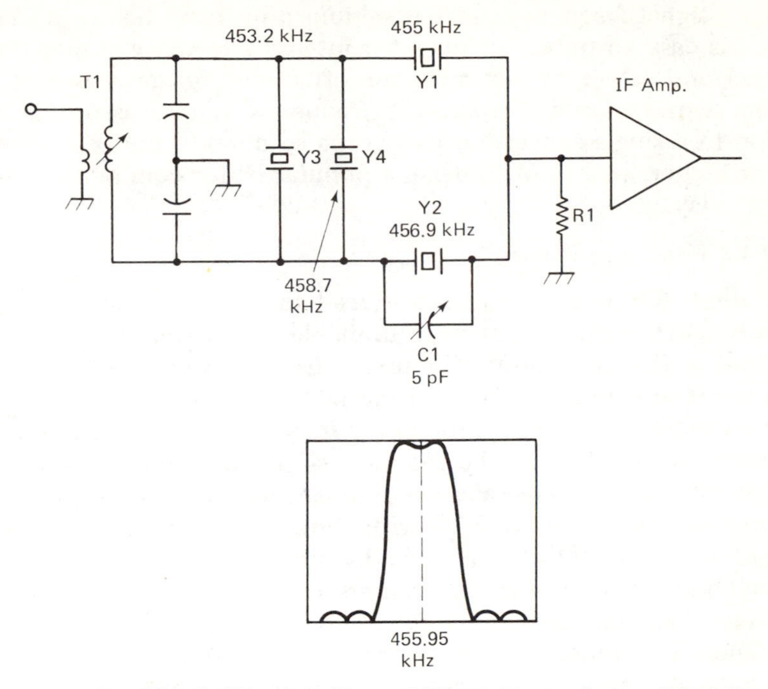

Figure 7.12 The half-lattice filter of Fig. 7.11 can be modified for improved skirt selectivity by adding shunt crystals *Y*3 and *Y*4, as shown.

at *Y*1. This will depend on the layout used in the filter and must be determined experimentally. *T*1 in Figs. 7.11 and 7.12 is tuned to the filter f_c. Comparison of the relative responses of the two filters will show the benefits of adding *Y*3 and *Y*4. *R*1 is chosen to provide a termination that yields minimum passband ripple. In simple filters of the kind seen in Figs. 7.11 and 7.12, the value of the load resistor is determined empirically.

Commercially made crystal lattice filters are available in a variety of center frequencies and bandwidths. They range from two to as many as ten poles, with a corresponding increase in skirt selectivity and cost. The major problem with multipole crystal filters is *crystal aging*. That is, a complex high-frequency filter can become misaligned as the crystals age and shift frequency, even slightly. Some manufacturers feel that the useful maximum number of poles is eight with respect to maintaining alignment in relation to time. The problem is not as critical when working with low-frequency crystals, since the quartz elements are relatively thick and stable. The disadvantage in using a low IF in a single-conversion HF-band receiver is the suppression of images (signals appearing above or below the IF, respective to the IF). For example, if the receiver front end did not have excellent selectivity, and if the IF was 455 kHz, image responses would be observed 455 kHz above or below the incoming signal frequency. This is seldom a problem below 5 MHz because it is easy to obtain sufficient front-end selectivity at medium frequencies and lower to minimize the effects of image responses. The problem worsens as the operating frequency is increased. The usual solution in a single-conversion receiver is to operate the IF system at 3 MHz or higher, with 9 MHz being a popular IF for communications receivers today.

7.4.2 Mechanical Filters

Collins Radio Company introduced the mechanical filter during the early 1950s. The filters were available also from a Japanese firm (Kokusai) in the mid 1960s. The useful frequency range of mechanical filters is 60 to 600 kHz. The limiting factor is the disc size. The discs become prohibitively large at the lower frequencies, and above 600 kHz they become too small to be practical. The numerous discs in a mechanical filter are metallic and are resonant at the IF. It is possible to obtain bandwidths down to 0.1% with a mechanical filter. For example, a 455-kHz filter could exhibit a bandwidth of only 45.5 Hz. If a coupling wire is passed through the centers of several discs, the fractional bandwidth can be made as great as 10% of the center frequency.

Figure 7.13 illustrates how a mechanical filter is structured. As the signal energy is routed through the input transducer, it is changed to

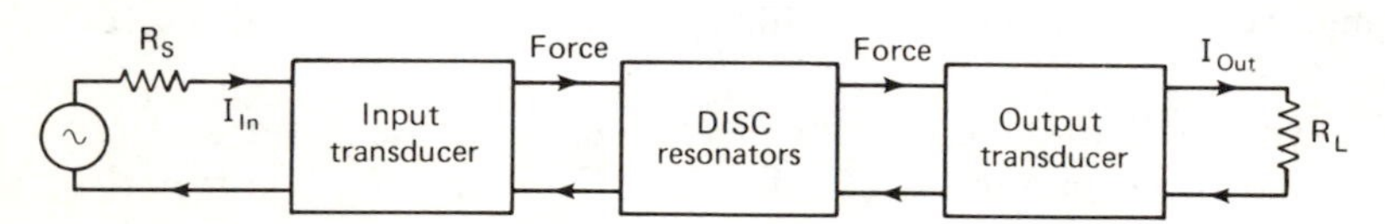

Figure 7.13 How a mechanical filter depends on electrical and mechanical energy to operate.

mechanical energy. This energy is passed through the resonant discs, and the undesired energy is filtered out. Finally, the mechanical energy is passed through the output transducer and converted back to electrical energy. The transducers serve also as terminations for the filter, since they reflect the source and load impedances to the mechanical section of the filter.

The insertion-loss range for mechanical filters is from 2 to 12 dB, depending upon the particular filter being used. Most mechanical filters have a characteristic impedance of 1000 to 2000 Ω. Resonating capacitors are required external to the filters for the purpose of ripple reduction. The manufacturer provides data concerning the capacitance needed to resonate the transducers. Generally, the value will be from 350 to 1100 pF.

A practical example of a circuit in which a mechanical filter is used appears in Fig. 7.14. $R1$ is selected for a value that is equivalent to

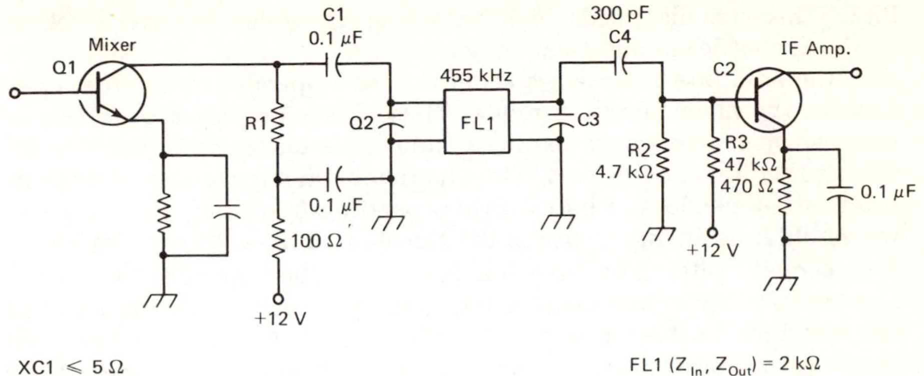

XC1 $\leqslant$ 5 Ω
C2, C3 : Per Mfg's. Specs.
R1 = FL1 (Z_{In})

FL1 (Z_{In}, Z_{Out}) = 2 kΩ
CH : Select to Provide FL1 (Z_{Out}) Match

$$XC4 = Z_{Out\,(FL1)} - (Z_{In\,(Q2)/R2})$$

$$\therefore XC4 = 2000 - \frac{1}{1000} + \frac{1}{4700} = 1175\ \Omega$$

$$C4 = \frac{1}{2\pi \times f_{mHz}} \times XC4 = \frac{1}{6.28 \times 0.455 \times 1175}$$

$$= \frac{1}{3357.4} = 0.000298\ \mu F$$

Figure 7.14 Typical arrangement for using a mechanical filter. *C*2 and *C*3 are used to resonate the filter input and output, thereby minimizing ripple. *C*4 matches the filter to its load, as indicated by the accompanying equations.

the filter impedance. Assuming our filter has an impedance of 2000 Ω, *R*1 would be 2000 Ω or the nearest standard value, such as 2200 Ω. *C*1 must have a capacitive reactance of 5 or less (0.07 µF or greater at 455 kHz). *C*2 and *C*3 are the resonating capacitors for the filter, as specified for the particular filter being used.

*FL*1 is terminated by *Q*2. The resistance it looks into is the parallel equivalent value of the transistor input impedance and *R*2. In effect, *R*3 is part of the determination, but since its value is much higher than the other two components mentioned, it can be ignored. The net resistance computes to 825 Ω, assuming a 100-Ω input impedance for *Q*2.

The filter needs to be terminated in 2000 Ω rather than 825 Ω. To satisfy this requirement, we can choose a *C*4 value that provides the difference between 825 and 2000 Ω, or 1175 Ω. This requires an X_c of 1175 Ω, which results in a capacitor value of 298 pF. The nearest standard value is 300 pF, which we find specified in Fig. 7.14 for *C*4.

7.4.3 Ceramic Filters

Ceramic filters are somewhat less expensive than the mechanical and crystal filters discussed earlier in this chapter. They make use of the piezoelectric effect in lead zirconate titanate ceramic material, or

PZT. This ceramic yields high values of Q while ensuring excellent coupling coefficient and frequency stability.

Ceramic filters contain a small disc that operates on the first overtone of the radial vibration mode. These filters provide a bandpass response when used singly or in cascade. This makes them suitable for use as IF filters in receivers. The characteristic impedances of ceramic filters are dependent in part on the operating frequency. For example, we would note an impedance in the range of 1000 to 2000 Ω for a 455-kHz ceramic filter. The insertion loss would be dependent upon the number of filter elements used, but generally ranges between 3 and 10 dB. Matching to the active parts of the receiver circuit can be handled easily with transformers. Ceramic filters are suitable in ladder-filter arrangements and can be designed by following image-parameter procedures.

Figure 7.15 shows a practical circuit in which a single-pole ceramic bandpass filter is used. It is coupled to IF amplifiers $Q1$ and $Q2$ by

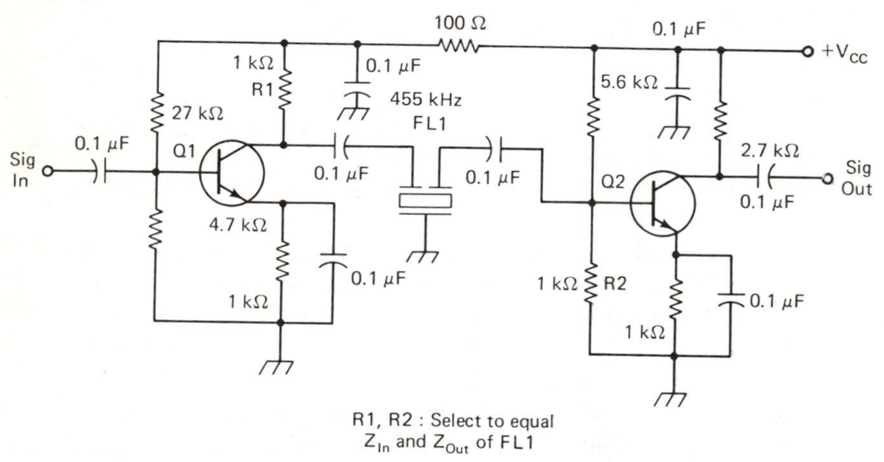

Figure 7.15 Piezo filters are often used where low cost is a consideration and when high orders of selectivity are not required.

means of 0.1-μF capacitors. Tuned transformers are often used for coupling these filters to the source and load impedances. $R1$ and $R2$ are selected for a value that equals the characteristic impedance of the filter. The skirt selectivity of a single-section ceramic filter is relatively broad. A more complex filter would be required in the interest of narrowing the passband characteristic.

A ceramic filter can be used as an emitter bypass, as seen in Fig. 7.16. It has a fairly broad and asymmetrical response, as shown by the accompanying curve. The actual response approaches that of a low-pass filter. Generally, ceramic filters find their greatest application in receivers that have a fairly broad bandpass characteristic, such as in FM receivers intended for the communications industry. Multipole filters

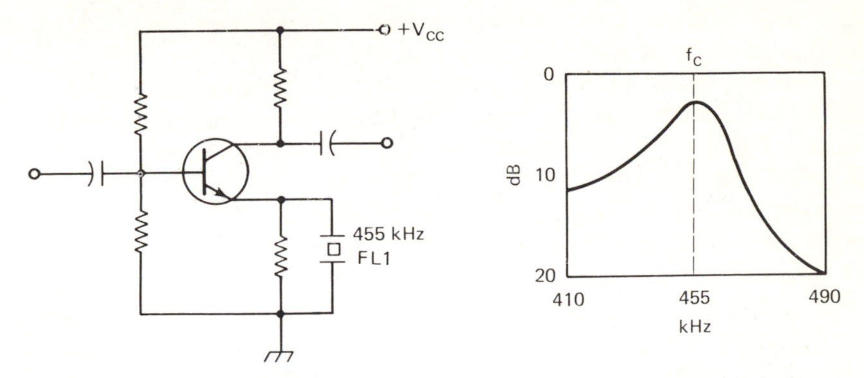

Figure 7.16 A piezo filter can be used as an emitter bypass to provide additional IF selectivity, as shown.

containing ceramic discs can be fabricated for use in SSB receivers when reduced cost is a major consideration. The inherent stability of ceramic filters makes them well suited to mobile applications.

7.4.4 Filter Switching

It becomes necessary in some design work to utilize two or more IF filters in the interest of having suitable bandpass characteristics for various communications modes. In this regard, we might design a receiver for the reception of AM, FM, SSB, and CW, thereby necessitating the inclusion of four IF filters. Mechancial switching can be cumbersome, and it may introduce stray coupling around the filter in use, spoiling its rejection capability: it is essential that the input and output ports of a filter be well isolated from one another if we are to preserve the filter characteristics. In some instances it is even necessary to place a shield divider across the bottom of the filter to aid isolation. As the filter operating frequency is elevated, the problem becomes more pronounced.

Diode switching, as illustrated in Fig. 7.17, offers a simple solution to filter switching. *S*1 is shown in position for actuating the CW filter. In this mode *D*3 and *D*4 are saturated, but *D*1 and *D*2 are reverse biased to aid isolation of *FL*1 from *FL*2. This technique can be expanded to accommodate as many filters as necessary.

The resistors in the switching circuit need to be low enough in value to permit hard switching of the diodes, but must not be so low in value that they have a major effect on the filter terminations. If, for example, a 2000-Ω filter were used, the 2200-Ω resistors would have a serious shunting effect on the filter. In situations of that kind we can insert RF chokes (X_L equal to or greater than four times the filter impedance) between the diodes and the switching resistors. A 10-mH RF

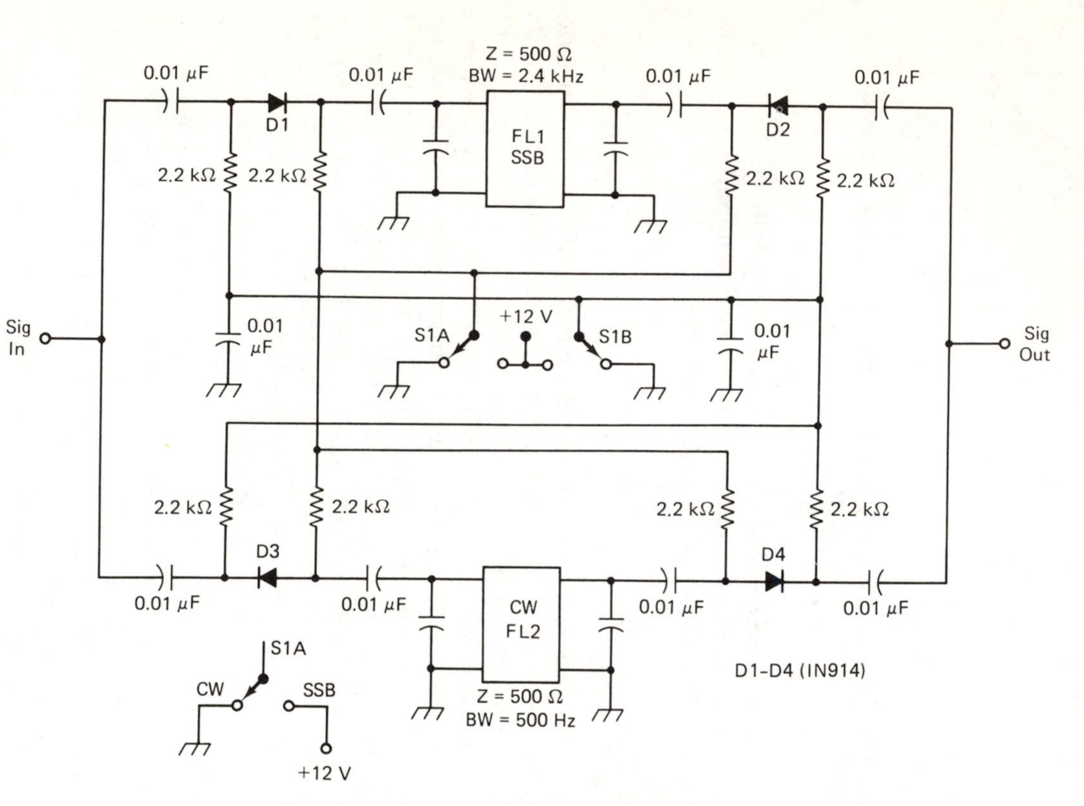

Figure 7.17 Silicon diodes can be used as switches when more than one IF filter is needed to accommodate additional modes of reception. Diode switching offers good input-output isolation for the filters and is less expensive than mechanical switching.

choke would be satisfactory for use with a 455-kHz filter, as one example.

7.5 AGC SYSTEMS

The main criteria for a receiver AGC system involves stability, attack time, decay time, the desired AGC control voltage, and the polarity of the AGC voltage. Some receiver parameters are subjective ones that we must decide upon in an effort to satisfy our individual preferences. The AGC attack and decay times are somewhat arbitrary in this respect. Generally, the attack time must be tailored to prevent a loud click in the receiver output when a strong signal actuates the AGC system. Fortunately, elements of R and C are all that are necessary in establishing the attack and decay times. The decay period is usually on the order of 1 to 2 seconds (s) for CW and SSB reception, although shorter or longer periods are preferred by some designers. A few modern communications receivers have a front-panel control that can be used to adjust the decay time of the AGC to as long as 5 s from an interval as short as a few microseconds.

Frequency stability is as important in an AGC system as it is in any other section of a receiver. The main area of difficulty is the AGC amplifier, owing to the fairly high gain that is required to permit light coupling to the RF or audio sampling point. The same stability precautions that apply to RF and IF amplifiers can be followed.

The AGC polarity will be governed by the types of IF amplifier devices used. We are not speaking here of positive and negative voltages, but rather about an increase in AGC voltage (or a decrease) as the incoming received signal becomes greater in amplitude. The CA3028A, for example, delivers maximum gain with maximum AGC voltage. The MC1590, conversely, provides maximum gain with minimum AGC voltage. It is in this context alone that we refer to the AGC polarity.

7.5.1 Simple AGC for MOSFETs

A simple but practical AGC circuit that is suitable for use with dual-gate MOSFETs in an IF amplifier is given in Fig. 7.18. It is an adaptation of a circuit designed by the author for use with CA3028As in an IF amplifier. Georges Ricaud made changes in the op-amp circuit and added three 1N914 diodes so that the system could be used to provide AGC to an IF amplifier that contained 40673s. The diodes serve to drop the AGC voltage to the desired maximum level for gate 2 of 40673s or 3N211s. A preamplifier precedes the AGC rectifier ($D1$) to ensure full AGC voltage at the output of $U1$.

$C1$ is selected to provide the desired AGC decay time. $R1$ can be made variable for the purpose of changing the AGC characteristics.

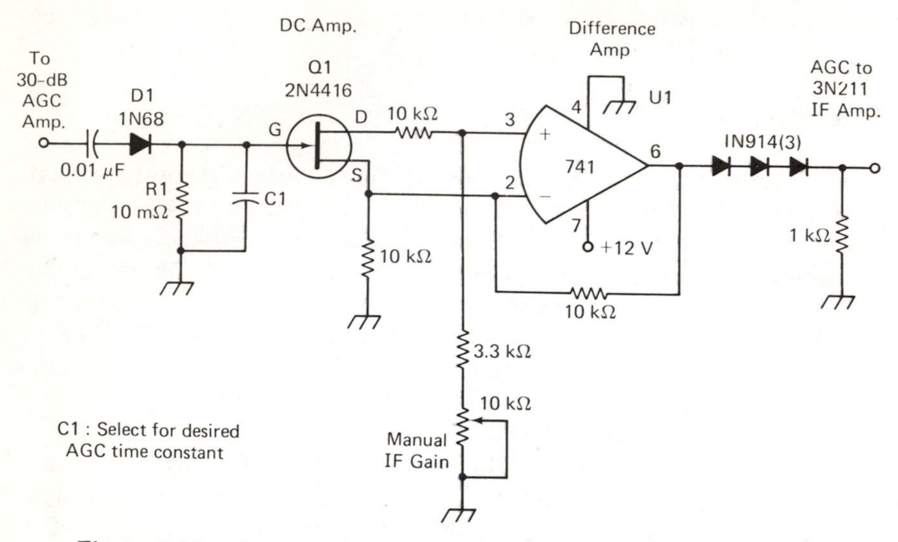

Figure 7.18 Simple IF control circuit for amplifiers that use AGC. This practical circuit is for use with dual-gate MOSFET IF amplifiers.

As the received signal increases in amplitude, *D*1 delivers additional dc voltage to the gate of *Q*1. The *Q*1 change in current is sampled by difference amplifier *U*1, thereby setting the output level of the op-amp to produce AGC voltage. To prevent undue loading on the circuit to which the AGC amplifier is connected (sampling point), very light coupling is used. A CA3028A or MC1350P IC would serve well as the AGC amplifier ahead of *D*1.

7.5.2 AGC System Using Discrete Devices

The practical circuit in Fig. 7.19 shows how an AGC system can be built from discrete transistors. It is also designed for use with dual-gate MOSFETs. The attack time is set for approximately 50 ms. Decay time is 5 s with the component vlaues shown. *C*1 can be made smaller in value if a shorter decay is desired.

*Q*1 operates as a tuned amplifier at the receiver IF. Various diodes are used in the circuit to establish references and voltage drops for the AGC line and the S-meter circuit. This circuit operates smoothly, and the signal-strength meter provides a reasonably linear response.

7.5.3 Other AGC Circuits

Figure 7.20 depicts two practical AGC control circuits for use with IC types of IF amplifiers. The version in Fig. 7.20(a) is set up for use with CA3028A IF amplifiers, which produce maximum gain with maximum AGC voltage applied. The circuitry ahead of the JFET dc

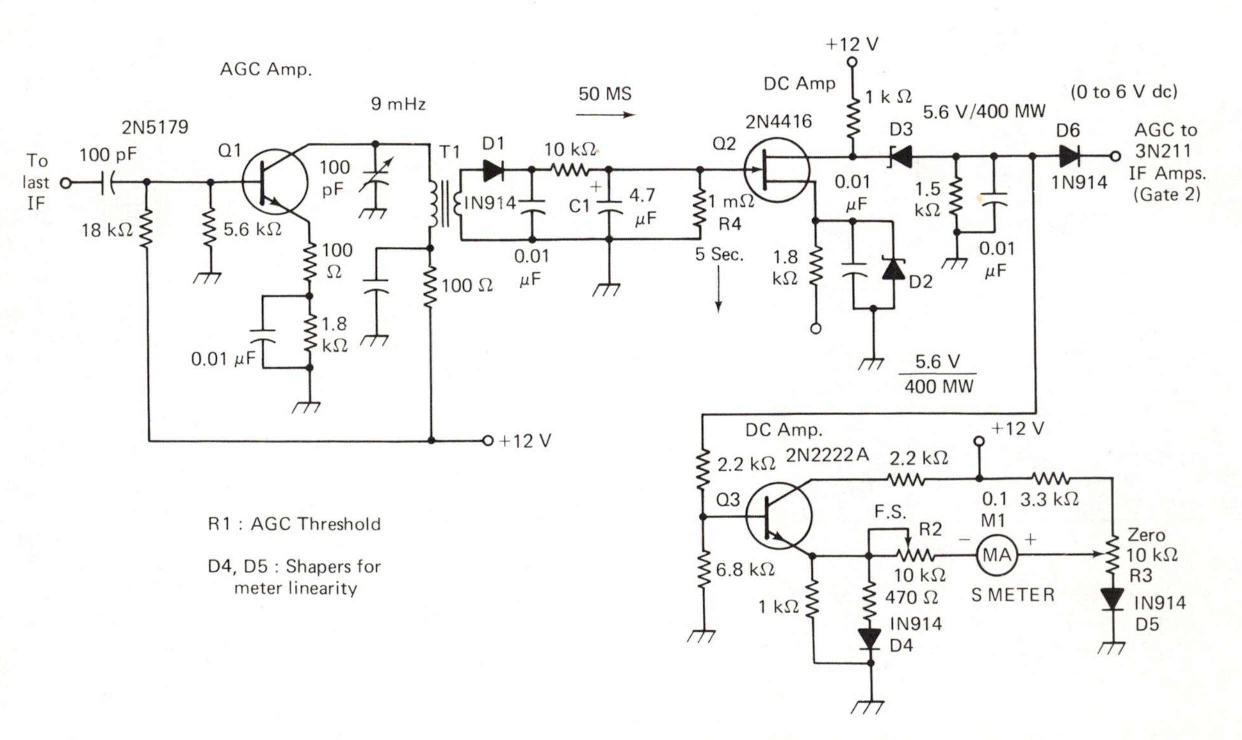

Figure 7.19 Complete AGC system for use with dual-gate MOSFET IF amplifiers. *R*1 is a threshold control that is set to provide the desired AGC initiation time respective to the incoming signal level. *D*4 and *D*5 are shaping diodes to aid S-meter linearity.

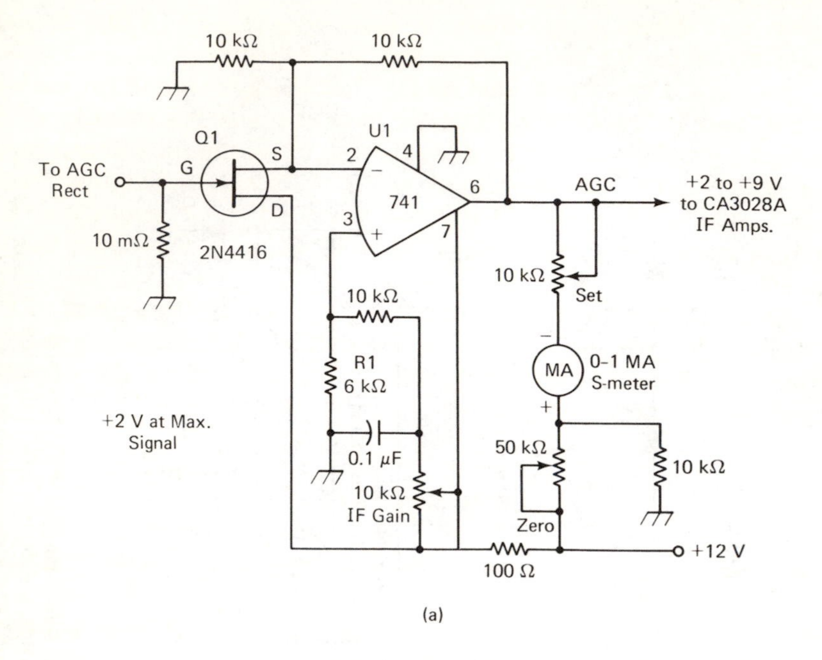

(a)

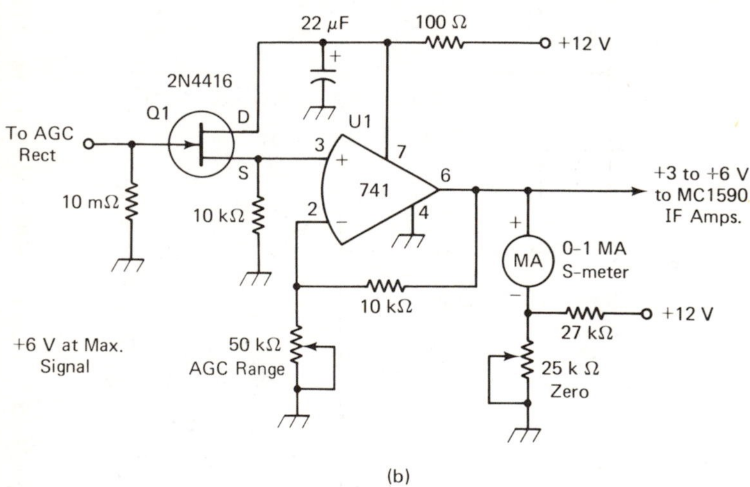

(b)

Figure 7.20 (a) Circuit designed to provide AGC to two CA3028A IF amplifiers; (b) a similar circuit, tailored for use with MC1590 or MC1350P IF amplifiers.

amplifier can be the same as that shown in Fig. 7.19. The S meter in this circuit does not yield a linear response. Rather, it approaches mid-scale at receiver input-signal levels less than 10 μV. From 10 to 10,000 μV, the meter deflects upward to full scale. The absolute value of $R1$ is chosen to provide an AGC voltage range of +2 to +9, maximum signal and zero signal, respectively. A small trim pot of 10,000 Ω could be substituted for $R1$ to permit on-the-nose adjustment.

The AGC circuit provided in Fig. 7.20(b) is intended for use with MC1590 and MC1350P ICs. Here we have changed the polarity of the op-amp to provide maximum AGC voltage at minimum receiver input-signal level. Either of the circuits in Fig. 7.20 can be adapted readily for use with other styles of ICs that are used in IF amplifiers. It is merely a matter of programming the AGC voltage range from the op-amp, $U1$.

As in the case with the circuit of Fig. 7.20(a), the circuit of Fig. 7.20(b) can be preceded by the AGC amplifier and detector presented in Fig. 7.19. Either of these AGC circuits can be used for RF- or audio-derived AGC systems.

BIBLIOGRAPHY

DeMaw, and W. Hayward, *Solid State Design for the Radio Amateur*, Chapter 5, ARRL, Inc., Newington, Conn., 1977.

Radio Corporation of America, *RCA Data File 382*, 1973, "Differential/Cascode Amplifiers," (covers CA3028A and similar ICs), 1973.

Rohde, U., "I-F Amplifier Design," *Ham Radio Magazine*, March 1977.

Trout, B., "A High-Gain Integrated-Circuit RF-IF Amplifier with Wide-Range AGC," *Motorola AN-513*, Phoenix, AZ.

INDEX